T0281338

Frontiers in Mathematics

Ivan Gavrilyuk
Volodymyr Makarov
Vitalii Vasylyk

Exponentially Convergent

Algorithms for

Abstract Differential

Equations

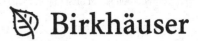 Birkhäuser

Ivan Gavrilyuk
Staatliche Studienakademie Thüringen -
Berufsakademie Eisenach
University of Cooperative Education
Am Wartenberg 2
99817 Eisenach
Germany
ipg@ba-eisenach.de

Volodymyr Makarov
Vitalii Vasylyk
Institute of Mathematics
National Academy of Sciences of Ukraine
3 Tereshchenkivska St.
01601 Kyiv- 4
Ukraine
makarov@imath.kiev.ua
vasylyk@imath.kiev.ua

2010 Mathematics Subject Classification: 65J10, 65M70, 35K90, 35L90

ISBN 978-3-0348-0118-8   e-ISBN 978-3-0348-0119-5
DOI 10.1007/978-3-0348-0119-5

Library of Congress Control Number: 2011932317

Cover design: deblik, Berlin

Printed on acid-free paper

Springer Basel AG is part of Springer Science+Business Media

www.birkhauser-science.com

Preface

The present book is based on the authors' works during recent years on constructing exponentially convergent algorithms and algorithms without accuracy saturation for operator equations and differential equations with operator coefficients in Banach space. It is addressed to mathematicians as well as engineering, graduate and post-graduate students who are interested in numerical analysis, differential equations and their applications. The authors have attempted to present the material based on both strong mathematical evidence and the maximum possible simplicity in describing algorithms and appropriate examples.

Our goal is to present accurate and efficient exponentially convergent methods for various problems, especially, for abstract differential equations with unbounded operator coefficients in Banach space. These equations can be considered as the meta-models of the systems of ordinary differential equations (ODE) and/or the partial differential equations (PDEs) appearing in various applied problems. The framework of functional analysis allows us to get rather general but, at the same time, present transparent mathematical results and algorithms which can be then applied to mathematical models of the real world. The problem class includes initial value problems (IVP) for first-order differential equations with a constant and variable unbounded operator coefficient in Banach space (the heat equation is a simple example), IVPs for first-order differential equations with a parameter dependent operator coefficient (parabolic partial differential equations with variable coefficients), IVPs for first-order differential equations with an operator coefficient possessing a variable domain (e.g. the heat equation with time-dependent boundary conditions), boundary value problems for second-order elliptic differential equations with an operator coefficient (e.g. the Laplace equation) as well as IVPs for second-order strongly damped differential equations. We present also exponentially convergent methods to IVP for first-order nonlinear differential equations with unbounded operator coefficients.

The efficiency of all proposed algorithms is demonstrated by numerical examples.

Unfortunately, it is almost impossible to provide a comprehensive bibliography representing the full set of publications related to the topics of the book. The list of publications is restricted to papers whose results are strongly necessary to follow the narrative. Thus we ask those who were unable to find a reference to their related publications to excuse us, even though those works would be appropriate.

Potential reader should have a basic knowledge of functional and numerical analysis and be familiar with elements of scientific computing. An attempt has been made to keep the presented algorithms understandable for graduate students in engineering, physics, economics and, of course, mathematics. The algorithms are suitable for such interactive, programmable and highly popular tools as MATLAB and Maple which have been extensively employed in our numerical examples.

The authors are indebted to many colleagues who have contributed to this work via fruitful discussions, but primarily to our families whose support is especially appreciated. We express our sincerest thanks to the German Research Foundation (Deutsche Forschungsgemeinschaft) for financial support. It has been a pleasure working with Dr. Barbara Hellriegel and with the Birkhäuser Basel publication staff.

Eisenach–Jena–Kyiv,   I. Gavrilyuk, V. Makarov, V. Vasylyk
April, 2011

Contents

Chapter 1

Introduction

The exponential function $y = e^x$ as well as the Euler number 'e' play fundamental roles in mathematics and possess many miraculous features expressing the fascination of nature. The exponential function is an important component of the solution of many equations and problems. For example, taking into account the commutative property of multiplication of numbers, the solution of the elementary algebraic equation

$$AX + XB = C, \quad A, B, C \in \mathbb{R}, \ A + B > 0 \tag{1.1}$$

can simply be found as

$$X = \frac{C}{A + B}. \tag{1.2}$$

At the same time, this solution can be represented in terms of the exponential function as

$$X = \int_0^\infty e^{-At} \cdot C \cdot e^{-Bt} dt.$$

Equation (1.1) is known as the Silvester equation and, when $A = B$, as the Lyapunov equation. If A, B, C are matrices or operators, the first formula can no longer be used (matrix multiplication is not commutative and we can not write $AX + XB = X(A+B)$!) but the second one remains true (under some assumptions on A and B). This means that the latter can, after proper discretization, be used in computations. In the case of a matrix or a bounded operator A, the exponential function e^{-At} (the operator exponential) can be defined, e.g., through the series

$$e^{-At} = \sum_{k=0}^\infty (-1)^k \frac{A^k t^k}{k!}. \tag{1.3}$$

There are also various other representations, e.g., it was shown in [2, 3, 20] that

$$
\begin{aligned}
\mathrm{e}^{-At} &= \mathrm{e}^{-\gamma t}\sum_{p=0}^{\infty}(-1)^{p}L_{p}^{(0)}(2\gamma t)T_{\gamma}^{p}(I+T_{\gamma})\\
&= \mathrm{e}^{-\gamma t}\sum_{p=0}^{\infty}(-1)^{p}\left[L_{p}^{(0)}(2\gamma t)-L_{p-1}^{(0)}(2\gamma t)\right]T_{\gamma}^{p},
\end{aligned}
\tag{1.4}
$$

where

$$
T_{\gamma}=(\gamma I-A)(\gamma I+A)^{-1}
$$

is the Cayley transform of the operator A, $L_{p}^{(0)}(t)$ are the Laguerre polynomials, and γ is an arbitrary positive real number. Another representation of e^{-At} is given by the Dunford-Cauchy integral

$$
\mathrm{e}^{-At}=\int_{\Gamma}\mathrm{e}^{-tz}(zI-A)^{-1}dz,
\tag{1.5}
$$

where the integration path Γ envelopes the spectrum of A. The last two representations can also be used for unbounded operators. In this case, the integral becomes improper.

The operator exponential in general is lacking some properties of the usual exponential function. For example, the definition (1.3) implies

$$
\begin{aligned}
\mathrm{e}^{-(A+B)t} &= I-(A+B)t+\frac{(A+B)^{2}}{2}t^{2}+\cdots,\\
\mathrm{e}^{-At}\mathrm{e}^{-Bt} &= \left(I-At+\frac{A^{2}t^{2}}{2}+\cdots\right)\left(I-Bt+\frac{B^{2}t^{2}}{2}+\cdots\right),\\
\mathrm{e}^{-(A+B)t}-\mathrm{e}^{-At}\mathrm{e}^{-Bt} &= [A,B]\frac{t^{2}}{2}+\cdots,
\end{aligned}
$$

where $[A,B]=BA-AB$ is the so-called commutator of the operators A and B. Therefore, if the well-known and important property of the exponential function

$$
\mathrm{e}^{-(A+B)t}=\mathrm{e}^{-At}\mathrm{e}^{-Bt},\quad \forall t
\tag{1.6}
$$

is valid for the operator exponential, then the commutator is equal to zero, i.e., the operators A and B commute, $AB=BA$. One can show that this condition is also sufficient for property (1.6).

The solution of the simple initial value problem

$$
\frac{dy}{dt}+Ay=0,\quad y(0)=b,\ \ A,b\in\mathbb{R}
\tag{1.7}
$$

for the ordinary differential equation is given by

$$
y(t)=\mathrm{e}^{-At}b.
\tag{1.8}
$$

One can say that the exponential function is a key to the solution of this problem or, in other words, the multiplication of the initial value by the exponential function is a "solution operation" (compare with "solution operator" for partial differential equations). The solution of the boundary value problem

$$\frac{d^2y}{dt^2} - Ay = 0, \quad y(0) = b, \ y(1) = c, \quad A, b, c \in \mathbb{R} \tag{1.9}$$

is given by

$$y(t) = -A^{-1} \sinh^{-1}\left(\sqrt{A}\right)\left(\sinh\left(\sqrt{A}t\right)c + \sinh\left(\sqrt{A}(1-t)\right)b\right),$$

where the functions involved are, once again, defined through the exponential function :

$$\sinh\left(\sqrt{A}t\right) = \frac{e^{\sqrt{A}t} - e^{-\sqrt{A}t}}{2}.$$

If A is a matrix or operator, equations of type (1.7), (1.9) can be considered as meta-models for various systems of ordinary or partial differential equations (respectively, ODE and PDE). For example, the initial value problem for the well-known heat equation

$$\frac{\partial u}{\partial t} - \frac{\partial^2 u}{\partial x^2} = 0,$$
$$u(0, x) = u_0(x), \ x \in (0, 1),$$
$$u(t, 0) = 0, \ u(t, 1) = 0$$

with a known function $u_0(x)$ and unknown function $u(t, x)$ can be written in the form (1.7) if we understand A as the operator defined by the differential expression $Av = -\frac{d^2 v}{dx^2}$ with domain $D(A)$ of twice differentiable functions $v(x)$ such that $v(0) = v(1) = 0$. The solution of (1.7) can then be formally written down in the form (1.8) with the so-called solution operator $T(t) = e^{-At}$ which maps the initial function $b = u_0(x)$ into the solution of problem (1.7). At this point the question arises: how we can compute an efficient approximation to this solution operator or to an element $e^{-At}b$?

When computing a function $f(t), \ t \in D$ from a space of functions X, the problem consists, in fact, of representing an element of, in general, an infinite-dimensional space X (continuum), through some n real numbers, i.e., by an element of $\mathbb{R}^n$. There are many ways to do that, but the mostly-used approximation method is to compute n approximate values of the function on a grid. Let us recall some facts from the theory of the so-called grid n-width, which characterize the goodness of the approximation of an arbitrary element from the function set X (see [5]). Let $X = W_p^r(M; I)$ with $M = (M_1, \ldots, M_d)$ and $r = (r_1, \ldots, r_d)$ be the class of anisotropic Sobolev spaces defined on the d-dimensional interval

$I = \prod_{j=1}^{d}[a_j, b_j]$. The constant $\rho = 1/(\sum_{j=1}^{d} r_j^{-1})$ defines the effective *smoothness of the class* and the constant $\mu = \prod_{j=1}^{d} M_j^{\rho/r_j}$ is the so-called *constant of the class*. With help of the n-width theory one can estimate the optimal number $n_\varepsilon^{(opt)}$ of the parameters (coordinates) needed to approximate an arbitrary function from this class with a given tolerance ε:

$$n_\varepsilon^{(opt)} \asymp \text{const}(\mu) \cdot \varepsilon^{-1/(\rho - 1/p)}.$$

It is also known that

$$n_\varepsilon^{(opt)} = \mathcal{O}\left((\log|\log\varepsilon|)^d |\log\varepsilon|^d\right)$$

numbers are needed to represent an analytical function of d variables with a tolerance ε.

We observe that, in general, $n_\varepsilon^{(opt)}$ grows exponentially as $d \to \infty$ (this phenomenon is known as "the curse of dimensionality"). It is of great practical importance since we need algorithms for representation of multidimensional functions, say with polynomial complexity in d, at least, for partial subclasses of functions.

The solutions or solution operators of many applied problems are analytic or piecewise analytic. If an algorithm for solution of such a problem uses a constant account of arithmetical operations per coordinate of the approximate solution vector, then the measure for the complexity of this algorithm (in the case $d = 1$) is of the order $\log\varepsilon^{-1}$. Let us suppose that an algorithm to find an analytic function $u(x)$ as the solution of an applied problem posits a vector y of n_a numbers where it holds that $\|u - y\| \le \phi(n_a)$. To arrive at the tolerance ε with the (asymptotically) optimal coordinate numbers $n_a \asymp \log\frac{1}{\varepsilon}$, the function $\phi(n_a)$ must be exponential. On the other hand, to be able to keep within the estimate $n_{opt} \asymp n_a \asymp \log\frac{1}{\varepsilon}$, the algorithm must possess a complexity $C = C(n_a)$ of the order $n_a \asymp \log\frac{1}{\varepsilon}$ with respect to the account of arithmetical operations (we call this *linear complexity*). If $C(n_a)$ is proportional to $n_a \log^\alpha n_a$ with α independent of n_a, we say that the algorithm possesses *almost linear complexity* .

Thus, an optimal algorithm for analytic solutions has to be exponentially convergent and possess a linear complexity such that the value $\log\varepsilon^{-1}$ is a near optimal complexity measure.

In the analysis above we have assumed that algorithms can operate with real numbers. Indeed, it is not the case; computer algorithms operate only with rational numbers, and we need a measure to estimate the goodness of a representation of functions from various classes by n *rational* numbers. This measure $H(\varepsilon; X)$ is called the ε-entropy of X (see [5], p. 245).

It is known that the method using numbers, say decimal numbers, of the maximal length N for representation of elements of a Banach space, possesses an accuracy ε if $N \ge H(\varepsilon; X)$, i.e., $H(\varepsilon; X)$ is the optimal length of numbers to represent X with a given tolerance ε.

For the anisotropic class W_p^r of functions, it holds that

$$H(\varepsilon; W_p^r) \asymp \left(\frac{\mu}{\varepsilon}\right)^{1/\rho}. \tag{1.10}$$

In the isotropic case, $r_1 = r_2 = \cdots = r_d = r$, we have $\rho = r/d$ and $H(\varepsilon; W_p^r) \asymp \left(\frac{\mu}{\varepsilon}\right)^{d/r}$, i.e., the length of the numbers increases exponentially with increasing dimension d. Thus, we again deal with the curse of dimensionality and the exponential function!

Let $I_0 = \{x \in \mathbb{R}^d : x = (x_1, x_2, \ldots, x_d), |x_j| \le 1\}$ and E_{r_j} be the domain in the complex plane $z_j = x_j + iy_j$ enveloped by an ellipse with focal points -1 and 1 whose sum of semi-axes is equal to $r_j > 1$ (the so-called Bernstein regularity ellipse: see below). We set $E_r = E_{r_1} \times \cdots \times E_{r_d}$. Let $X(E_r, I_0; M)$ be the compact of continuous on I_0 functions with the usual Chebyshev norm $\| \cdot \|_\infty$ which can be extended analytically into E_r and are bounded by a positive constant M.

For the class $A(E_r, I_0; M)$ of analytic functions, it holds [5], p. 262 that

$$H(\varepsilon; A(E_r, I_0; M)) = \frac{2}{(d+1)! \prod_j \log r_j} \log^{d+1} \frac{M}{\varepsilon} + \mathcal{O}\left(\left(\log \frac{M}{\varepsilon}\right)^d \log \log \frac{M}{\varepsilon}\right).$$

We see that, for analytic functions, the curse of dimensionality is somewhat weaker since in the isotropic case we have

$$H(\varepsilon; A(E_r, I_0; M)) \asymp \log^{d+1} \frac{M}{\varepsilon};$$

however, it remains still exponential in d.

Having based this point on the considerations above, we can derive the following requirement of a numerical algorithm of minimal complexity. It should possess as far as possible:

1) an exponential convergence rate,

2) a polynomial complexity in d.

We shall see that the exponential function and the operator exponential can be a key to fulfilling these aims within the framework of the present book. To achieve the second goal, a tensor-product approximation with an exponential operator exponential, property (1.6) in particular can be helpful as an important integral part (see Appendix).

Exponentially convergent algorithms were proposed recently for various problems. The corresponding analysis is often carried out in an abstract setting. This means that the initial value and boundary value problems of parabolic, hyperbolic and elliptic type are formulated as abstract differential equations with an operator coefficient A:

$$\frac{du}{dt} + Au = f(t), \quad t \in (0, T]; \quad u(0) = u_0, \tag{1.11}$$

$$\frac{d^2u}{dt^2} + Au = f(t), \quad t \in (0,T]; \quad u(0) = u_0, \quad u'(0) = u_{01}, \qquad (1.12)$$

$$\frac{d^2u}{dx^2} - Au = -f(x), \quad x \in (0,1); \quad u(0) = u_0, \quad u(1) = u_{01}. \qquad (1.13)$$

Here, A is the densely defined closed (unbounded) operator with domain $D(A)$ in a Banach space X, u_0, u_{01} are given vectors and $f(t)$ is a given vector-valued function. The operator A is, e.g., an elliptical differential operator. In particular, it can be the Laplace operator, i.e., in the simplest one-dimensional case, the second spatial derivative with appropriate boundary conditions:

$$D(A) := \{v \in H^2(0,1) : v(0) = 0, \ v(1) = 0\},$$
$$Av := -\frac{d^2v}{dy^2} \quad \text{for all } v \in D(A). \qquad (1.14)$$

Which are typical approaches to construct exponentially convergent algorithms for the operator differential equations above? One of the possibilities is to represent the solution of (1.11), (1.12) and (1.13) by using the corresponding solution operator (exponential etc.) families. For the governing equation (1.11), this representation reads as

$$u(t) = e^{-At}u_0 + \int_0^t e^{-A(t-\tau)}f(\tau)\,d\tau. \qquad (1.15)$$

The problems (1.12) and (1.13) yield analogous integral equations with the operator-cosine $\cos(t\sqrt{A})$ and operator-hyperbolic-sine $\sinh^{-1}\sqrt{A}\sinh(x\sqrt{A})$ family (normalized), respectively.

The majority of the exponentially convergent methods based on this approach require a suitable representation of the corresponding operator functions and its approximations. A convenient representation of the operator functions in Banach space X can be provided by an improper Dunford-Cauchy integral (see, e.g., [9]), e.g., for the operator exponential given by (1.5). By employing a parametrization of the integration path $z = z(\xi) \in \Gamma_I$, $\xi \in (-\infty, \infty)$ we get an improper integral of the type

$$e^{-At} = \frac{1}{2\pi i}\int_\Gamma e^{-tz}(zI - A)^{-1}dz = \frac{1}{2\pi i}\int_{-\infty}^{\infty}\mathcal{F}(t,\xi)d\xi \qquad (1.16)$$

with $\mathcal{F}(t,\xi) = e^{-tz(\xi)}(z(\xi)I - A)^{-1}z'(\xi)$.

The integral representation, e.g., (1.16), can then be discretized by a (possibly exponentially convergent) quadrature formula involving a short sum of resolvents. Such algorithms for linear homogeneous parabolic problems of the type (1.11) were proposed (probably first) in [23, 24] and later on in [16, 18, 25, 27, 59, 65]. These algorithms are based on a representation of the operator exponential $T(t) = e^{-At}$ by the improper Dunford-Cauchy integral along a path enveloping the spectrum of

A where a hyperbola containing the spectrum of A or a parabola as the integration path are used. The methods from [16, 23, 27, 30] use Sinc-quadratures [44, 62] and possess an exponential convergence rate. An exponential convergence rate for all $t \geq 0$ was proven in [17, 69] under assumptions that the initial function u_0 belongs to the domain of $D(A^\sigma)$ for some $\sigma > 1$, where the preliminary computation of $A^\sigma u_0$ is needed. Note that not all of these algorithms can be directly applied to inhomogeneous problems due to inefficient computation of the operator exponential at $t = 0$. In [43, 42] the exponentially convergent algorithms were used to invert the Laplace transform with a better (up to a logarithmical factor) convergence rate. However, these methods do not provide uniform convergence for all $t \geq 0$ and the parallel realization for different time values. In [27], a hyperbola as the integration path and proper modification of the resolvent were used. This made it possible to get the uniform and numerically stable exponential convergence rate with respect to $t \geq 0$ without preliminary computation of $A^\sigma u_0$. An exponentially convergent algorithm for the case of an operator family $A(t)$ depending on the parameter t was proposed in [26]. This algorithm uses an exponentially convergent algorithm for the operator exponential generation by a constant operator. Moreover, these algorithms inherit two levels of parallelism (with respect to various time points and with respect to the treatment of the summands in the quadrature sum) which was perhaps first observed in the paper [58] and, independently for exponentially convergent algorithms, in [16, 23, 24]. A parallel method for the numerical solution of an integro-differential equation with positive memory was described in [39]. The paper [12] deals with exponentially convergent algorithms for parabolic PDEs based on Runge-Kutta methods. The problem of constructing exponentially convergent approximations for operator sine and cosine functions is still open. The following exact representation for a solution of the operator cosine-function with generator A was proposed in [15, 22] (see also [25]):

$$C(t) \equiv \cos \sqrt{A}t = e^{-\delta t} \sum_{n=0}^{\infty} \left(L_n^{(0)}(t) - L_{n-1}^{(0)}(t) \right) \mathcal{U}_n,$$

which implies the next representation of the solution of problem (1.12),

$$
\begin{aligned}
x(t) &\equiv x(t; A) = (\cos \sqrt{A}t)x_0 \\
&= e^{-\delta t} \sum_{n=0}^{\infty} \left(L_n^{(0)}(t) - L_{n-1}^{(0)}(t) \right) u_n,
\end{aligned}
\tag{1.17}
$$

where δ is an arbitrary real number in $(-1/2, \infty)$ and $L_n^{(0)}(t)$ are the Laguerre polynomials. The sequences of vectors $\{u_n\}$ and of operators $\{\mathcal{U}_n\} \equiv \{\mathcal{U}_n(A)\}$ are defined by

$$
\begin{aligned}
u_{n+1} &= 2(A + \delta(\delta - 1)I)(A + (\delta - 1)^2 I)^{-1} u_n \\
&\quad - (A + \delta^2 I)(A + (\delta - 1)^2 I)^{-1} u_{n-1}, \quad n \geq 1, \\
u_0 &= x_0, \quad u_1 = (A + \delta(\delta - 1)I)(A + (\delta - 1)^2 I)^{-1} x_0
\end{aligned}
\tag{1.18}
$$

and

$$\begin{aligned}
\mathcal{U}_{n+1} = {} & 2(A + \delta(\delta - 1)I)(A + (\delta - 1)^2 I)^{-1}\mathcal{U}_n \\
& - (A + \delta^2 I)(A + (\delta - 1)^2 I)^{-1}\mathcal{U}_{n-1}, \quad n \geq 1, \\
\mathcal{U}_0 = {} & I, \quad \mathcal{U}_1 = (A + \delta(\delta - 1)I)(A + (\delta - 1)^2 I)^{-1}
\end{aligned} \tag{1.19}$$

without using $\sqrt{A}$. The operator A was supposed to be strongly P-positive, i.e., its spectrum is enveloped by a parabola in the right half-plane and the resolvent on and outside of the parabola satisfies

$$\|(sI - A)^{-1}\| \leq \frac{M}{1 + \sqrt{z}}.$$

As an approximation of the exact solution, one can use the truncated series consisting of the first N summands. This approximation does not possess accuracy saturation, i.e., the accuracy is of the order $\mathcal{O}(N^{-\sigma})$, $\sigma > 0$ as $N \to \infty$, σ characterizes the smoothness of the initial data and this accuracy decreases exponentially provided that the initial data are analytical.

In the present book we develop our results in constructing exponentially convergent methods for the aforementioned problems. We are sure that this book can not clarify all the remarkable properties and applications of the exponential function. We would therefore urge the reader to continue our excursion into the wonderful world of applications of the exponential function.

Chapter 2

Preliminaries

In this chapter we briefly describe some relevant basic results on interpolation, quadratures, estimation of operators and representation of operator-valued functions using the Dunford-Cauchy integral.

2.1 Interpolation of functions

Interpolation of functions by polynomials is one of the basic ideas in designing numerical algorithms. Let Π_n be a class of polynomials of degree less than or equal to n and a smooth function $u(x)$ be defined on (a, b). The interpolation polynomial $P_n(x) = P_n(u; x)$ for the function $u(x)$ satisfies the conditions:

- $P_n(u; x) \in \Pi_n$,
- for $n + 1$ various points in (a, b), $x_0, x_1, \ldots, x_n \in (a, b)$,

$$P_n(u; x_i) = u(x_i), \quad i = \overline{0, n}.$$

The interpolation polynomial is unique and can be written down in various forms, e.g., in the form by Lagrange (discovered by Joseph Louis Lagrange in 1795)

$$P_n(u; x) = L_n(u, x) = \sum_{i=0}^{n} u(x_i) L_{i,n}(x), \qquad (2.1)$$

where

$$L_{i,n}(x) = \prod_{\substack{j=0, \\ j \neq i}}^{n} \frac{x - x_i}{x_i - x_j} = \frac{q_{n+1}(x)}{q'_{n+1}(x_i)(x - x_i)}, \quad i = \overline{0, n}; \qquad (2.2)$$

$$q_{n+1}(x) = (x - x_0) \cdots (x - x_n)$$

are called the Lagrange fundamental polynomials. It is easy to see that

$$L_{i,n}(x_j) = \begin{cases} 1, j = i, \\ 0, j \neq i. \end{cases}$$

Defining the k-th *divided difference* recursively by

$$u[x_i, x_{i+1}, \ldots, x_{i+k}] = \frac{u[x_{i+1}, \ldots, x_{i+k}] - u[x_i, x_{i+1}, \ldots, x_{i+k-1}]}{x_k - x_i}, \quad k = 1, 2, \ldots, n;$$

$$u[x_i] = u(x_i),$$

the Newton form of the interpolation polynomial (for equidistant nodes, found in the 1670s by Isaak Newton) is given by

$$P_n(u; x) = u[x_0] + \sum_{i=1}^{n} (x - x_0) \cdots (x - x_i) u[x_0, x_1, \ldots, x_i].$$

Introducing the auxiliary quantities

$$\lambda_k^{(n)} = \prod_{i=1, i \neq k}^{n} \frac{1}{x_k - x_i},$$

the interpolation polynomial can be written down as the following *barycentric formula*:

$$P_n(u; x) = \frac{\sum_{k=0}^{n} \frac{\lambda_k^{(n)}}{x - x_k} u(x_k)}{\sum_{i=0}^{n} \frac{\lambda_i^{(n)}}{x - x_i}}$$

which possesses good computational properties [47].

One can hope that, if the grid of interpolation nodes covers the interpolation interval from dense to dense, then the interpolation polynomial converges to the function. In order to describe this process let us introduce the following infinite triangular array of interpolation knots:

$$\mathcal{X} = \begin{pmatrix} x_{0,0} & 0 & \cdots & 0 & \cdots \\ x_{0,1} & x_{1,1} & \cdots & 0 & \cdots \\ \cdot & \cdot & \cdots & \cdot & \cdots \\ x_{0,n} & x_{1,n} & \cdots & x_{n,n} & \cdots \\ \cdot & \cdot & \cdots & \cdot & \cdots \end{pmatrix},$$

where the $(n + 1)$-th row contains the zeros of $q_{n+1}(x) = (x - x_{0,n}) \cdots (x - x_{n,n})$. Now we can associate to the array $\mathcal{X}$ (or to $q_{n+1}(x)$) a sequence of Lagrange interpolation polynomials $\{L_n(\mathcal{X}, u)\}_{n \in \mathbb{N}}$, defined by

$$L_n(\mathcal{X}, u)(x_{n,k}) = L_n(\mathcal{X}, u; x_{n,k}) = u(x_{n,k}), \quad k = 0, 1, \ldots, n.$$

These polynomials can be written in form (2.1) by putting $x_{i,n}$ instead of x_i and $L_{i,n}(\mathcal{X};x)$ instead of $L_{i,n}(x)$. For a given interpolation array $\mathcal{X}$ we define a sequence of linear operators $L_n(\mathcal{X}) : C[-1,1] \rightarrow \Pi_n, n = 0,1,\dots$ such that $L_n(\mathcal{X})u = L_n(\mathcal{X},u)$. The sequence $\{L_n(\mathcal{X})\}_{n\in\mathbb{N}}$ defines an interpolatory process and, for each $n \in \mathbb{N}$, the operator $L_n(\mathcal{X})$ is a projector onto Π_n. The main question is the convergence of $L_n(\mathcal{X},u;x) \rightarrow u(x)$ when $n \rightarrow \infty$ and the corresponding convergence rate is a norm. It is easy to compare the *interpolation error* $\varepsilon_I(u) = \|u - L_n(\mathcal{X},u)\|$ with the error of best approximation $E_n(u)$ given by

$$E_n(u) = \min_{p\in\Pi_n} \|u - p\| = \|u - P^*\|,$$

where $P^* \in \Pi_n$ is the polynomial of the best uniform approximation to u in the corresponding norm.

Let us consider, for example, the space $C^0 = C[-1,1]$ of all continuous functions equipped with the uniform norm

$$\|u\| = \|u\|_\infty = \max_{|x|\leq 1} |u(x)|.$$

We have

$$
\begin{aligned}
|u(x) - \dot{L}_n(\mathcal{X},u;x)| &\leq |u(x) - P^*(x)| + |P^* - L_n(\mathcal{X},u;x)| \\
&\leq E_n(u) + |L_n(\mathcal{X},u - P^*;x)| \\
&\leq (1 + \lambda_n(\mathcal{X};x))E_n(u),
\end{aligned}
\tag{2.3}
$$

where $\lambda_n(\mathcal{X};x) = \sum\limits_{k=1}^{n}|L_{k,n}(\mathcal{X};x)|$ is the *Lebesgue function*. Further, this yields

$$\varepsilon_I(u) \leq \|u(x) - L_n(\mathcal{X},u)\| \leq (1 + \Lambda_n(\mathcal{X}))E_n(u), \tag{2.4}$$

where $\Lambda_n(\mathcal{X}) = \|\lambda_n(\mathcal{X};x)\|$ is called the *Lebesgue constant*. Thus, the first factor which influences the convergence of the interpolation process is this constant which depends only on the knots distribution. In 1914 Faber proved that for the uniform norm

$$\Lambda_n \geq \frac{1}{12} \log n, n \geq 1. \tag{2.5}$$

Besides, one can prove that:

1) there exist continuous functions for which the interpolation process diverges,

2) there exists an optimal nodes distribution, for which there exists a constant $C \neq C(n)$ such that $\|\Lambda_n(\mathcal{X};x)\|_\infty \leq C \log n$.

The interpolation points which produce the smallest value Λ_n^* of all Λ_n are not known, but Bernstein in 1954 proved that

$$\Lambda_n^* = \frac{2}{\pi} \log n + O(1).$$

An algorithm to compute the optimal system of nodes can be found in [47].

The optimal system of nodes is known analytically only for some special cases. For example, if the $(n+1)$-th row of the array of interpolation knots $\mathcal{X} = \mathcal{X}(v^{(\alpha,\beta)})$ consists of zeros of the n-th Jakobi orthogonal polynomials with weight $v^{(\alpha,\beta)}(x) = (1-x)^\alpha (1+x)^\beta$, the following classical result due to Szegö holds [47, p. 248]:

$$\|\Lambda_n(\mathcal{X};x)\|_\infty \asymp \begin{cases} \log n, & \text{if } -1 < \alpha, \beta \leq -1/2, \\ n^{\max\{\alpha,\beta\}+1/2}, & \text{otherwise.} \end{cases} \qquad (2.6)$$

Often it is necessary to use the interpolation including the ends of the segment $[-1,1]$. In this case it is natural to use the so-called Chebyshev-Gauss-Lobatto interpolation with the knots being the zeros of the polynomial $(1-x^2)U_{n-1}(x) = (1-x^2)T_n'(x)$, where $U_{n-1}(x)$ is the Chebyshev polynomial of the second kind [47, p. 249]. These knots are also known as *practical abscissas or Clenshaw's abscissas*. The Lagrange fundamental polynomials in this case can be written in the form

$$L_{i,n}(\eta) = \frac{(1-\eta^2)T_n'(\eta)}{(\eta - x_i)\frac{d}{d\eta}[(1-\eta^2)T_n'(\eta)]_{\eta = x_i}}, \quad i = 0,1,\ldots,n,$$

$$x_j = \cos\left(\pi(n-j)/n\right), \quad j = 0,1,\ldots,n.$$

If $\mathcal{X}$ is the system of the practical abscissas or Clenshaw's abscissas, then the corresponding Lebesgue constant is optimal: $\|\Lambda_n(\mathcal{X};x)\|_\infty \asymp \log n$. There are other Jakobi polynomials $J_n^{(\alpha,\beta)}$ whose roots build an optimal system [47, p. 254].

For the array $\mathcal{E}$ of equidistant nodes on $[-1,1]$ it holds that (see [47]) $\|L_n(\mathcal{E})\|_\infty \asymp 2^n/(en\log n), n \to \infty$.

It is clear that the Lebesque constant depends on the choice of the norm. For $\mathcal{X}$ being zeros of the $(n+1)$-th orthogonal polynomial on $[a,b]$ with a weight $\rho(x) \geq 0$, it holds that [63]

$$\|u - L_n(\mathcal{X},u)\|_{2,\rho} = \left(\int_a^b \rho(x)\big(u(x) - L_n(\mathcal{X},u;x)\big)^2 dx\right)^{1/2}$$

$$\leq 2\left(\int_a^b \rho(x)dx\right)^{1/2} E_n(u).$$

In the case $(a,b) \equiv (-1,1)$ *Chebyshev nodes*, i.e., the zeros of a Chebyshev polynomial of the first kind $T_{n+1}(x) = \cos((n+1)\arccos(x))$, are often used as the interpolation points. These polynomials may be defined recursively by

$$T_0(w) = 1, \quad T_1(w) = w,$$
$$T_{n+1}(w) = 2wT_n(w) - T_{n-1}(w), \quad n = 1,2,\ldots.$$

In the case of such *Chebyshev interpolation* it can be shown that Λ_n grows at most logarithmically in n, more precisely,

$$\Lambda_n \leq \frac{2}{\pi} \log n + 1.$$

The second factor which affects the convergence of the interpolation process is the error of best approximation $E_n(u)$ depending only on the smoothness of the function $u(x)$. For example, if all derivatives of $u(x)$ up to the order r belong to $L_p(a,b)$ and $\|u^{(r)}\|_p \leq M$, then this dependence for the uniform norm is polynomial with respect to n:

$$E_n(u) \leq A_r M n^{-r} \tag{2.7}$$

with a constant A_r independent of n, M.

A very interesting and important function class is the class of analytic functions. Before we come to the estimate for $E_n(u)$ for this function class, let us recall some basic facts about these functions.

In the complex plane $\mathbb{C}$, we introduce the circular ring

$$\mathcal{R}_\rho := \{z \in \mathbb{C} : 1/\rho < |z| < \rho\} \quad \text{with } \rho > 1$$

and the class $\mathcal{A}_\rho$ of functions $f : \mathbb{C} \to \mathbb{C}$ which are analytic and bounded by $M > 0$ in $\mathcal{R}_\rho$ with $\rho > 1$ and set

$$C_n := \frac{1}{2\pi} \int_0^{2\pi} f(e^{i\theta}) e^{in\theta} d\theta, \quad n = 0, \pm 1, \pm 2, \ldots. \tag{2.8}$$

The Laurent theorem asserts that if $f \in \mathcal{A}_\rho$, then $f(z) = \sum_{n=-\infty}^{\infty} C_n z^n$ for all $z \in \mathcal{R}_\rho$, where the series (the Laurent series) converges to $f(z)$ for all $z \in \mathcal{R}_\rho$. Moreover $|C_n| \leq M/\rho^{|n|}$, and, for all $\theta \in [0, 2\pi]$ and arbitrary integer m,

$$\left| f(e^{i\theta}) - \sum_{n=-m}^{m} C_n e^{in\theta} \right| \leq \frac{2M}{\rho - 1} \rho^{-m}. \tag{2.9}$$

By $\mathcal{E}_\rho = \mathcal{E}_\rho(B)$ with the reference interval $B := [-1, 1]$, we denote the *Bernstein regularity ellipse* (with foci at $w = \pm 1$ and the sum of semi-axes equal to $\rho > 1$),

$$\mathcal{E}_\rho := \{w \in \mathbb{C} : |w - 1| + |w + 1| \leq \rho + \rho^{-1}\},$$

or

$$\mathcal{E}_\rho = \left\{ w \in \mathbb{C} : w = \frac{1}{2} \left(\rho e^{i\varphi} + \frac{1}{\rho} e^{-i\phi} \right) \right\}$$

$$= \left\{ (x, y) : \frac{x^2}{a^2} + \frac{y^2}{b^2} = 1, \ a = \frac{1}{2} \left(\rho + \frac{1}{\rho} \right), \ b = \frac{1}{2} \left(\rho - \frac{1}{\rho} \right) \right\}.$$

It can be seen that for the Chebyshev polynomials, there holds

$$T_n(w) = \frac{1}{2}(z^n + z^{-n}) \tag{2.10}$$

with $w = \frac{1}{2}(z + \frac{1}{z})$.

If F is analytic and bounded by M in $\mathcal{E}_\rho$ (with $\rho > 1$), then the expansion (Chebyshev series)

$$F(w) = C_0 + 2\sum_{n=1}^{\infty} C_n T_n(w), \tag{2.11}$$

holds for all $w \in \mathcal{E}_\rho$, where

$$C_n = \frac{1}{\pi}\int_{-1}^{1} \frac{F(w)T_n(w)}{\sqrt{1-w^2}}dw.$$

Moreover, $|C_n| \le M/\rho^n$ and for $w \in B$ and for $m = 1, 2, 3, \ldots$,

$$\left|F(w) - C_0 - 2\sum_{n=1}^{m} C_n T_n(w)\right| \le \frac{2M}{\rho-1}\rho^{-m}, \quad w \in B. \tag{2.12}$$

Let $\mathcal{A}_{\rho,s} := \{f \in \mathcal{A}_\rho : C_{-n} = C_n\}$, then each $f \in \mathcal{A}_{\rho,s}$ has a representation (cf. (2.11))

$$f(z) = C_0 + \sum_{n=1}^{\infty} C_n(z^n + z^{-n}), \quad z \in \mathcal{R}_\rho. \tag{2.13}$$

Furthermore, from (2.13) it follows that $f(1/z) = f(z), \ z \in \mathcal{R}_\rho$.

Let us apply the mapping $w(z) = \frac{1}{2}(z + \frac{1}{z})$, which satisfies $w(1/z) = w(z)$. It is a conformal transform of $\{\xi \in \mathcal{R}_\rho : |\xi| > 1\}$ onto $\mathcal{E}_\rho$ as well as of $\{\xi \in \mathcal{R}_\rho : |\xi| < 1\}$ onto $\mathcal{E}_\rho$ (but not $\mathcal{R}_\rho$ onto $\mathcal{E}_\rho$!). It provides a one-to-one correspondence of functions F that are analytic and bounded by M in $\mathcal{E}_\rho$ with functions f in $\mathcal{A}_{\rho,s}$.

Since under this mapping we have (2.10), it follows that if f defined by (2.13) is in $\mathcal{A}_{\rho,s}$, then the corresponding transformed function $F(w) = f(z(w))$ that is analytic and bounded by M in $\mathcal{E}_\rho$ is given by (2.13).

Now the result (2.12) follows directly from the Laurent theorem.

The next assertion gives an error estimate for the interpolation polynomial for analytic functions: Let $u \in C^\infty[-1,1]$ have an analytic extension to $\mathcal{E}_\rho$ bounded by $M > 0$ in $\mathcal{E}_\rho$ (with $\rho > 1$). Then we have

$$\|u - L_n u\|_{\infty,I} \le (1 + \Lambda_n)\frac{2M}{\rho-1}\rho^{-n}, \quad n \in \mathbb{N}_{\ge 1}. \tag{2.14}$$

Actually, due to (2.12) one obtains for the best polynomial approximations to u on $[-1,1]$,

$$\min_{v \in \mathcal{P}_N} \|u - v\|_{\infty,B} \le \frac{2M}{\rho-1}\rho^{-N}. \tag{2.15}$$

Note that the interpolation operator L_n is a projection, that is, for all $v \in \Pi_n$ we have $L_n v = v$. Then applying the triangle inequality with $v \in \Pi_n$,

$$\|u - L_n u\|_{\infty,B} = \|u - v - L_n(u - v)\|_{\infty,B} \leq (1 + \Lambda_n)\|u - v\|_{\infty,B}$$

completes the proof.

To present some more estimates for the interpolation error (see, e.g., [7]), we define weighted L^p-norms as follows:

$$\|u\|_{L_w^p(-1,1)} = \left(\int_{-1}^{1} |u(x)|^p w(x) dx \right)^{1/p} , \quad \text{for} \quad 1 \leq p < \infty,$$

$$\|u\|_{L_w^\infty(-1,1)} = \sup_{-1 \leq x \leq 1} |u(x)|,$$

and weighted Sobolev spaces

$$\|u\|_{H_w^m(-1,1)} = \left(\sum_{k=0}^{m} \|u^{(k)}(x)\|_{L_w^2(-1,1)}^2 \right)^{1/2},$$

with seminorms

$$|u|_{H_w^{m;N}(-1,1)} = \left(\sum_{k=min(m,N+1)}^{m} \|u^{(k)}(x)\|_{L_w^2(-1,1)}^2 \right)^{1/2}.$$

For the truncation error we have

$$\|u - P_n(u, \cdot)\|_{L_w^2(-1,1)} \leq Cn^{-m}|u|_{H_w^{m;n}(-1,1)}, \tag{2.16}$$

for all $u \in H_w^m(-1,1)$ with $m \geq 0$. For higher-order Sobolev norms the estimate is

$$\|u - P_n(u, \cdot)\|_{L_w^l(-1,1)} \leq Cn^{2l-m}|u|_{H_w^{m;n}(-1,1)}, \tag{2.17}$$

for all $u \in H_w^m(-1,1)$ with $m \geq 1$ and $1 \leq l \leq m$.

The estimates (2.16) and (2.17) remain true for the Chebyshev-Gauss-Lobatto interpolation too.

Sometimes it is useful to use other basis functions different from x^i. Let $\Pi_{g,n}$ be a class of generalized polynomials $P_{g,n}(u, x) = \sum_{i=0}^{n} c_i g_i(x)$ of degree less than or equal to n in a basis $\{g_i(x)\}_{i=0,...,n}$ (instead of $\{x^i\}_{i=0,...,n}$). Then, the generalized interpolation polynomial for a continuous function $u(x)$ is defined by the following conditions:

- $P_{g,n}(u, x) \in \Pi_{g,n}$,

- for $n + 1$ various points in (a, b), $x_0, x_1, \ldots, x_n \in (a, b)$,

$$P_{g,n}(u, x_i) = u(x_i), \quad i = \overline{0, n}.$$

The conditions on the basis function (they should build the so-called Chebyshev system) under which this problem possesses the unique solution can be found in [11].

One of the possible choices is the Sinc basis. To illustrate an approximation technique using the Sinc basis we use the notation (see [62]):

$$D_d = \{z \in \mathbb{C} : -\infty < \Re z < \infty, \ |\Im z| < d\},$$

where $D_d(\epsilon)$ is defined for $0 < \epsilon < 1$ by

$$D_d(\epsilon) = \{z \in \mathbb{C} : |\Re z| < 1/\epsilon, \ |\Im z| < d(1 - \epsilon)\}.$$

Also we introduce the space $\mathbf{H}^p(D_d)$ of all functions such that for each $f \in \mathbf{H}^p(D_d)$ there holds $\|f\|_{\mathbf{H}^p(D_d)} < \infty$ with

$$\|f\|_{\mathbf{H}^p(D_d)} = \begin{cases} \lim\limits_{\epsilon \to 0} \left(\int_{\partial D_d(\epsilon)} |f(z)|^p |dz| \right)^{1/p} & \text{if } 1 \le p < \infty, \\ \lim\limits_{\epsilon \to 0} \sup\limits_{z \in \partial D_d(\epsilon)} |f(z)| & \text{if } p = \infty. \end{cases}$$

Let

$$S(k,h)(x) = \frac{\sin\left[\pi(x - kh)/h\right]}{\pi(x - kh)/h} \tag{2.18}$$

be the k-th Sinc function with step size h, evaluated in x. Given $f \in \mathbf{H}^p(D_d)$, $h > 0$ and a positive integer N, let us use the notation

$$C(f,h) = \sum_{k=-\infty}^{\infty} f(kh)S(k,h),$$

$$C_N(f,h) = \sum_{k=-N}^{N} f(kh)S(k,h),$$

$$E(f,h) = f - C(f,h) \quad E_N(f,h) = f - C_N(f,h).$$

V.A. Kotelnikov (1933), J. Wittacker (1935) and C.E. Shannon(1948) have proved that band-limited signals can be exactly reconstructed via their sampling values or, mathematically speaking, if the support of the Fourier transform $\hat{f}(\omega)$ of a continuous function $f(t)$ is included in $[-\pi/h, \pi/h]$, then

$$f(t) = \sum_{n=-\infty}^{\infty} f(nh)S(n,h)(t) \equiv C(f,h)(t).$$

The truncated sum $C_N(f,h)$ is the generalized interpolation polynomial with the basis functions $S(k,h)(t)$ for the function $f(t)$, i.e., $C_N(f,h)(kh) = f(kh), k = -N, -N+1, \ldots, N$.

The following error representation theorem holds true [62]

Theorem 2.1. *Let $f \in \mathbf{H}^p(D_d)$ with $1 \leq p \leq \infty$. Then for $z \in D_d$,*

$$E(f, h)(z) = f(z) - C(f, h)(z)$$
$$= \frac{\sin(\pi z/h)}{2\pi i} \int_{-\infty}^{\infty} \left\{ \frac{f(t - id^-)}{(t - z - id)\sin[\pi(t - id)/h]} - \frac{f(t + id^-)}{(t - z + id)\sin[\pi(t + id)/h]} \right\} dt.$$

This theorem yields the following error estimate for functions with an exponential decay.

Theorem 2.2. *Let the conditions of Theorem 2.1 be fulfilled and f satisfy*

$$|f(x)| \leq ce^{-\alpha|x|}, \quad x \in (-\infty, \infty),$$

with positive constants c, α; then, taking

$$h = \left(\frac{\pi d}{\alpha N} \right)^{1/2},$$

there exists a positive number C_1, depending only on f, d, α, and y, such that for $s = 2$ or $s = \infty$,

$$\|E_N(f, h)\|_s \leq C_1 N^{(1-1/s)/2} \exp\left\{ -\left(\frac{\pi\alpha}{d} \right)^{1/2} (d - |y|) N^{1/2} \right\}.$$

2.2 Exponentially convergent quadrature rule

The main purpose of numerical analysis and scientific computing is to develop efficient and accurate methods to compute approximations to quantities that are difficult or impossible to obtain by analytic means. One of the most frequently encountered problems in constructing numerical methods is the approximation of integrals that arise. A very simple idea to compute an integral

$$I = I(u) = \int_a^b u(x)dx \tag{2.19}$$

is to replace the integrand $u(x)$ by an aproximation $\tilde{u}(x)$ so that the integral $I(u) = \int_a^b \tilde{u}(x)dx$ is "computable". If we choose the interpolant

$$L_n(u, x) = \sum_{i=0}^{n} u(x_i) L_{i,n}(x) \tag{2.20}$$

as $\tilde{u}(x)$, then we obtain the *quadrature formula*

$$I = I(u) = \int_a^b u(x)dx \approx \mathcal{Q}_n u = \sum_{i=0}^{n} c_{i,n} u(x_i) \tag{2.21}$$

with the coefficients $c_{i,n} = \int_a^b L_{i,n}(x)dx$ which can be calculated analytically. Such quadrature formulas are called *interpolation quadrature formulas*. Their error can be estimated by

$$\varepsilon_Q(u) = \left| \int_a^b [u(x) - [\mathcal{I}_N u](x)]dx \right| \leq (b-a)\varepsilon_I(u) \qquad (2.22)$$

and has the same order with respect to N as the interpolation error.

We mentioned in the introduction that, for a given tolerance ε, an exponential convergence rate provides algorithms of optimal or low complexity. Such algorithms were developed for approximation of various types of integrals based on Sinc-quadratures. Let us consider briefly the problem of approximation of the integral (2.19) with $a = -\infty$, $b = \infty$ which we will use in this monograph. For a given $f \in \mathbf{H}^p(D_d)$, $h > 0$ (see [62]) and positive integer N, let us use the notation

$$I(f) = \int_{\mathbb{R}} f(x)dx, \quad T_N(f,h) = h \sum_{k=-N}^{N} f(kh), \quad T(f,h) = h \sum_{k=-\infty}^{\infty} f(kh),$$

$$\eta_N(f,h) = I(f) - T_N(f,h), \quad \eta(f,h) = I(f) - T(f,h).$$

The following theorem holds true [62].

Theorem 2.3. *Let $f \in \mathbf{H}^1(D_d)$. Then*

$$\eta(f,h) = \frac{i}{2} \int_{-\infty}^{\infty} \left\{ \frac{f(t - id^-)e^{-\pi(d+it)/h}}{\sin[\pi(t - id)/h]} - \frac{f(t + id^-)e^{-\pi(d-it)/h}}{\sin[\pi(t + id)/h]} \right\} dt.$$

Moreover,

$$|\eta(f,h)| \leq \frac{e^{-\pi d/h}}{2\sin(\pi d/h)} \|f\|_{\mathbf{H}^1(D_d)}.$$

If in addition, f satisfies

$$|f(x)| \leq ce^{-\alpha|x|}, \quad x \in (-\infty, \infty),$$

with positive constants c, α, then, taking

$$h = \left(\frac{2\pi d}{\alpha N} \right)^{1/2},$$

we obtain

$$|\eta_N(f,h)| \leq c_1 e^{-(2\pi d\alpha N)^{1/2}},$$

with c_1 depending only on f, d and α.

This result can be extended to the case of vector-valued functions using the Bochner integral [1, 71]. So, we have a quadrature rule that possesses exponential convergence for the class of analytic functions with exponential decay on $\pm\infty$. Using this quadrature, we develop in the next chapters new methods for solution of some problems.

2.3 Estimates of the resolvent through fractional powers of strongly positive operators

Let A be a densely defined strongly positive (sectorial) operator in a Banach space X with domain $D(A)$, i.e., its spectrum $\Sigma(A)$ lies in the sector

$$\Sigma = \{z = a_0 + re^{i\theta} : r \in [0,\infty), |\theta| < \varphi < \frac{\pi}{2}\} \tag{2.23}$$

and on its boundary Γ_Σ and outside the sector the following estimate for the resolvent holds true:

$$\|(zI - A)^{-1}\| \le \frac{M}{1 + |z|} \tag{2.24}$$

with some positive constant M (compare with [20, 35, 54, 60]). The angle φ is called the spectral angle of the operator A. A practically important example of strongly positive operators in $X = L_p(\Omega)$, $0 < p < \infty$ represents a strongly elliptic partial differential operator [14, 15, 16, 20, 26, 54, 56] where the parameters a_0, φ of the sector Σ are defined by its coefficients.

For an initial vector $u_0 \in D(A^{m+1})$ it holds that

$$\sum_{k=1}^{m+1} \frac{A^{k-1}u_0}{z^k} + \frac{1}{z^{m+1}}(zI - A)^{-1}A^{m+1}u_0 = (zI - A)^{-1}u_0 \tag{2.25}$$

This equality together with

$$A^{-(m+1)}v = \frac{1}{2\pi i} \int_{\Gamma_I} z^{-(m+1)}(zI - A)^{-1}v dz \tag{2.26}$$

by setting $v = A^{m+1}u_0$ yields the representation

$$u_0 = A^{-(m+1)}A^{m+1}u_0 = \frac{1}{2\pi i} \int_{\Gamma_I} z^{-(m+1)}(zI - A)^{-1}A^{m+1}u_0 dz$$

$$= \int_{\Gamma_I} \left[(zI - A)^{-1} - \sum_{k=1}^{m+1} \frac{A^{k-1}}{z^k}\right] u_0 dz \tag{2.27}$$

with an integration path Γ_I situated in the right half-plane and enveloping Γ_Σ. Let us estimate the norm of the first integrand in (2.27) as a function of $|z|$ under the assumption $u_0 \in D(A^{m+\alpha})$, $m \in \mathbb{N}$, $\alpha \in [0,1]$. Since the operator A is strongly positive it holds on and outside the integration path

$$\|(zI - A)^{-1}w\| \le \frac{M}{1 + |z|}\|w\|,$$

$$\|A(zI - A)^{-1}w\| \le (1 + M)\|w\|. \tag{2.28}$$

These estimates yield (see, e.g., Theorem 4 from [35])

$$\|A^{1-\alpha}(zI-A)^{-1}w\| \leq K\|A(zI-A)^{-1}w\|^{1-\alpha}\|(zI-A)^{-1}w\|^{\alpha}, \qquad (2.29)$$

where the constant K depends on α and M only. This inequality, taking into account (2.28), implies

$$\|A^{1-\alpha}(zI-A)^{-1}\| \leq \frac{K(1+M)}{(1+|z|)^{\alpha}}, \qquad \alpha \in [0,1] \qquad (2.30)$$

which leads to the estimate

$$\left\|\left[(zI-A)^{-1} - \frac{1}{z}I\right]u_0\right\| = \frac{1}{|z|}\|A^{1-\alpha}(zI-A)^{-1}A^{\alpha}u_0\| \qquad (2.31)$$

$$\leq \frac{(1+M)K}{|z|(1+|z|)^{\alpha}}\|A^{\alpha}u_0\|, \qquad \forall \alpha \in [0,1], \ u_0 \in D(A^{\alpha}).$$

This estimate can be easily generalized to

$$\left\|\left[(zI-A)^{-1} - \sum_{k=1}^{m+1}\frac{A^{k-1}}{z^k}\right]u_0\right\| = \left\|\frac{1}{z^{m+1}}(zI-A)^{-1}A^{m+1}u_0\right\|$$

$$= \frac{1}{|z|^{m+1}}\|A^{1-\alpha}(zI-A)^{-1}A^{m+\alpha}u_0\| \leq \frac{1}{|z|^{m+1}}\frac{(1+M)K}{(1+|z|)^{\alpha}}\|A^{m+\alpha}u_0\|, \qquad (2.32)$$

$$\forall \alpha \in [0,1], \ u_0 \in D(A^{m+\alpha}).$$

Thus, we get the following result which we will need below.

Theorem 2.4. *Let $u_0 \in D(A^{m+\alpha})$ for some $m \in \mathbb{N}$ and $\alpha \in [0,1]$; then the estimate* (2.32) *holds true.*

2.4   Integration path

There are many possibilities to define and to approximate functions of an operator A. Let Γ be the boundary of a domain Σ in the complex plane containing the spectrum of A and $\tilde{f}(z)$ be an analytical function in Σ; then the Dunford-Cauchy integral

$$\tilde{f}(A) = \frac{1}{2\pi i}\int_{\Gamma}\tilde{f}(z)(zI-A)^{-1}dz \qquad (2.33)$$

defines an operator-valued function $\tilde{f}(A)$ of A provided that the integral converges (see, e.g., [9]).

By a parametrizing of $\Gamma = \{z = \xi(s) + i\eta(s) : s \in (-\infty, \infty)\}$, one can translate integral (2.33) into the integral

$$\tilde{f}(A) = \int_{-\infty}^{\infty} F(s)ds \qquad (2.34)$$

with

$$F(s) = \frac{1}{2\pi i} \tilde{f}(z)(zI - A)^{-1} z'(s). \qquad (2.35)$$

Choosing various integration paths and using then various quadrature formulas, one can obtain various approximations of $\tilde{f}(A)$ with desired properties (see, e.g., [16, 17, 18, 20, 26, 30]). A comparison of some integration contours can be found in [70].

It was shown in [8, 15, 16, 23, 56] that the spectrum of a strongly elliptic operator in a Hilbert space that lies in a domain enveloped by a parabola defined by the coefficients of the operator and the resolvent on and outside of the parabola is estimated by

$$\|(zI - A)^{-1}\| \leq \frac{M}{1 + \sqrt{|z|}} \qquad (2.36)$$

with some positive constant M. Such operators are called strongly P-positive operators. The paper [56] contains also examples of differential operators which are strongly P-positive in such genuine Banach spaces as $L_1(0,1)$ or $L_\infty(0,1)$. One of the natural choices of the integration path for these operators is a parabola which does not intersect the spectral parabola containing the spectrum of the operator.

Let

$$\Gamma_0 = \{z = \xi - i\eta : \ \xi = a_0 \eta^2 + b_0, \ a_0 > 0, \ b_0 > 0, \ \eta \in (-\infty, \infty)\} \qquad (2.37)$$

be the spectral parabola enveloping the spectrum of the operator A. In [16, 23], it was shown how one can define the coefficients of an integration parabola by the coefficients of the spectral parabola so that the integrand in (2.34) can be analytically extended into a symmetric strip D_d of a width $2d$ around the real axes, however this choice was rather complicated.

Below we propose another (simpler) method to define the integration parabola through the spectral one.

We have to choose an integration parabola

$$\Gamma_I = \{z = \xi - i\eta : \ \xi = a_I \eta^2 + b_I, \ a_I > 0, \ b_I > 0, \ \eta \in (-\infty, \infty)\} \qquad (2.38)$$

so that its top lies in $(0, b_0)$ and its opening is greater than the one of the spectral parabola, i.e., $a_I < a_0$. Moreover, by changing η to $\eta + i\nu$, the set of parabolas

$$\begin{aligned}
\Gamma(\nu) = \{z = \xi - i\eta : \\
\xi = a_I \eta^2 + b_I - a_I \nu^2 + \nu - i\eta(1 - 2a_I\nu), \eta \in (-\infty, \infty)\} \\
= \{z = \xi - i\tilde{\eta} : \ \xi = \frac{a_I}{(1 - 2a_I\nu)^2}\tilde{\eta}^2 + b_I - a_I\nu^2 + \nu, \\
\tilde{\eta} = (1 - 2a_I\nu)\eta \in (-\infty, \infty)\},
\end{aligned} \qquad (2.39)$$

for $|\nu| < d$ must lie outside of the spectral parabola (only in this case one can guarantee that the resolvent of A remains bounded). Note, that the substitution

$\tilde{\eta} = (1 - 2a_I\nu)\eta$ must be nonsingular for all $|\nu| < d$, which yields $a_I < 1/(2d)$. We choose d so that the top of the integration parabola coincides with the top of the spectral one and the opening of the integration parabola is greater than the opening of the spectral parabola for $\nu = d$. For $\nu = -d$, we demand that the integration parabola lies outside of the spectral parabola and its top lies at the origin. Thus, it must be

$$\begin{cases} \dfrac{a_I}{(1 - 2a_I d)^2} = a_0, \\ b_I - a_I d^2 + d = b_0, \\ b_I - a_I d^2 - d = 0. \end{cases} \tag{2.40}$$

It follows immediately from the last two equations that $2d = b_0$. From the first equation

$$4d^2 a_0 a_I^2 - a_I(1 + 4a_0 d) + a_0 = 0, \tag{2.41}$$

after inserting $d = b_0/2$ we get

$$a_I = \frac{1 + 2a_0 b_0 \pm \sqrt{1 + 4a_0 b_0}}{2a_0 b_0} \tag{2.42}$$

but only the root

$$a_I = \frac{1 + 2a_0 b_0 - \sqrt{1 + 4a_0 b_0}}{2a_0 b_0} = \frac{2a_0}{1 + 2a_0 b_0 + \sqrt{1 + 4a_0 b_0}} \tag{2.43}$$

satisfies the condition $a_I < 1/(2d) = 1/b_0$. Thus, the parameters of the integration parabola from which the integrand can be analytically extended into the strip D_d of the width

$$d = b_0/2 \tag{2.44}$$

are

$$a_I = \frac{1 + 2a_0 b_0 - \sqrt{1 + 4a_0 b_0}}{2a_0 b_0} = \frac{2a_0}{1 + 2a_0 b_0 + \sqrt{1 + 4a_0 b_0}},$$
$$b_I = \frac{a_I b_0^2}{4} + \frac{b_0}{2}. \tag{2.45}$$

In the next section we will use an integration hyperbola which envelopes the spectral parabola and provides uniform approximations of the operator exponential including $t = 0$.

Chapter 3

The first-order equations

This chapter deals with problems associated with differential equations of the first order with an unbounded operator coefficient A in Banach space. The operator-valued function e^{-tA} (generated by A) plays an important role for these equations. This function is called also an operator exponential. In section 3.1, we present exponentially convergent algorithms for an operator exponential generated by a strongly positive operator A in a Banach space X. These algorithms are based on representations of e^{-tA} by the Dunford-Cauchy integral along various paths enveloping the spectrum of A combined with a proper quadrature involving a short sum of resolvents where the choice of the integration path effects dramatically desired features of the algorithms. Parabola and hyperbola as integration paths are analyzed and scales of estimates in dependence on the smoothness of initial data, i.e., on the initial vector and the inhomogeneous right-hand side, are obtained. One of the algorithms possesses an exponential convergence rate for the operator exponential e^{-At} for all $t \geq 0$ including the initial point. This makes it possible to construct an exponentially convergent algorithm for inhomogeneous initial value problems. The algorithm admits a natural parallelization. It turns out that the resolvent must be modified in order to obtain numerically stable algorithms near the initial point. The efficiency of the proposed method is demonstrated by numerical examples.

Section 3.2 is devoted to numerical approximation of the Cauchy problem for a first-order differential equation in Banach and Hilbert space with an operator coefficient $A(t)$ depending on the parameter t. We propose a discretization method with a high parallelism level and without accuracy saturation, i.e., the accuracy adapts automatically to the smoothness of the solution. For analytical solutions, the convergence rate is exponential. These results can be considered as a development of parallel approximations of the operator exponential e^{-tA} with a constant operator A possessing the exponential accuracy.

In Section 3.3, we consider the Cauchy problem for a first-order nonlinear equation with an operator coefficient in a Banach space. An exponentially convergent approximation to the solution is proposed. The algorithm is based on the equivalent Volterra integral formulation including the operator exponential generated by the operator coefficient. The operator exponential is represented by the Dunford-Cauchy integral along a hyperbola enveloping the spectrum of the operator coefficient and, thereafter, the involved integrals are approximated by using Chebyshev interpolation and an appropriate Sinc quadrature. Numerical examples are given which confirm theoretical results.

Section 3.4 is devoted to the first-order differential equation with an operator coefficient possessing a variable domain. A suitable abstract setting of the initial value problem for such an equation is introduced for the case when both the unbounded operator coefficient and its domain depend on the differentiation variable t. A new exponentially convergent algorithm is proposed. The algorithm is based on a generalization of the Duhamel integral for vector-valued functions. The Duhamel-like technique makes it possible to transform the initial problem to an integral equation and then approximate its solution with exponential accuracy. The theoretical results are confirmed by examples associated with heat transfer boundary value problems.

3.1 Exponentially convergent algorithms for the operator exponential with applications to inhomogeneous problems in Banach spaces

We consider the problem

$$\frac{du(t)}{dt} + Au(t) = f(t), \quad u(0) = u_0, \tag{3.1}$$

where A is a strongly positive operator in a Banach space X, $u_0 \in X$ is a given vector, $f(t)$ is a given, and $u(t)$ is the unknown vector-valued function. A simple example of a partial differential equation associated with the abstract setting (3.1) is the classical inhomogeneous heat equation

$$\frac{\partial u(t,x)}{\partial t} - \frac{\partial^2 u(t,x)}{\partial x^2} = f(t,x)$$

with the corresponding boundary and initial conditions, where the operator A is defined by

$$D(A) = \{v \in H^2(0,1): \ v(0) = 0, \ v(1) = 0\},$$

$$Av = -\frac{d^2 v}{dx^2} \quad \text{for all } v \in D(A).$$

The homogeneous equation

$$\frac{dT(t)}{dt} + AT(t) = 0, \quad T(0) = I, \tag{3.2}$$

where I is the identity operator and $T(t)$ is an operator-valued function that defines the semi-group of bounded operators $T(t) = \mathrm{e}^{-At}$ generated by A. This operator is called also the operator exponential or the solution operator of the homogeneous equation (3.1). Given the solution operator, the initial vector u_0, and the right-hand side $f(t)$, the solution of the homogeneous initial value problem (3.1) can be represented by

$$u(t) = u_h(t) = T(t)u_0 = \mathrm{e}^{-At}u_0 \tag{3.3}$$

and the solution of the inhomogeneous problem by

$$u(t) = \mathrm{e}^{-At}u_0 + u_p(t) \tag{3.4}$$

with

$$u_p(t) = \int_0^t \mathrm{e}^{-A(t-\xi)} f(\xi)d\xi. \tag{3.5}$$

We can see that an efficient approximation of the operator exponential is needed to get an efficient discretization of both (3.3) and (3.4). Further, having in mind a discretization of the second summand in (3.4) by a quadrature sum, we need an efficient approximation of the operator exponential for all $t \geq 0$ including the point $t = 0$.

3.1.1 New algorithm with integration along a parabola and a scale of estimates

Let A be a strongly P-positive operator and

$$u_0 \in D(A^\alpha), \quad \alpha > 0. \tag{3.6}$$

In this case due to (2.32) with $m = 0$, we have

$$\|[(zI - A)^{-1} - \frac{1}{z}I]u_0\| = \|\frac{1}{z}(zI - A)^{-1}Au_0\|$$
$$= \frac{1}{|z|}\|A^{1-\alpha}(zI - A)^{-1}A^\alpha u_0\| \leq \frac{1}{|z|}\|A^{1-\alpha}(zI - A)^{-1}\|\|A^\alpha u_0\|. \tag{3.7}$$

The resolvent of the strongly P-positive operator is bounded on and outside the spectral parabola, more precisely, we have

$$\|(zI - A)^{-1}w\| \leq \frac{M}{1 + \sqrt{|z|}}\|w\|,$$
$$\|A(zI - A)^{-1}w\| \leq \left(1 + \frac{M|z|}{1 + \sqrt{|z|}}\right)\|w\| \leq (1 + M\sqrt{|z|})\|w\|. \tag{3.8}$$

We suppose that our operator A is strongly positive (note that a strongly elliptic operator is both strongly P-positive [15] and strongly positive). We can use

Theorem 4 from [35] and get

$$\|A^{1-\alpha}(zI - A)^{-1}w\| \le K(\alpha)(1 + M\sqrt{|z|})^{1-\alpha} \left(\frac{M}{1 + \sqrt{|z|}}\right)^{\alpha} \|w\|$$

$$\le \max(1, M)K(\alpha)\frac{\|w\|}{(1 + \sqrt{|z|})^{2\alpha - 1}}$$

(3.9)

with a constant $K(\alpha)$ independent of α, where $K(1) = K(0) = 1$. The last inequality and (3.6) implies

$$\|[(zI - A)^{-1} - \frac{1}{z}I]u_0\| \le \max(1, M)K(\alpha)\frac{\|A^{\alpha}u_0\|}{|z|^{\alpha + 1/2}}$$

(3.10)

which justifies the following representation of the solution of the homogeneous problem (3.1) for the integration path above:

$$u(t) = e^{-At}u_0 = \frac{1}{2\pi i}\int_{\Gamma_I} e^{-tz}(zI - A)^{-1}u_0 dz$$

$$= \frac{1}{2\pi i}\int_{\Gamma_I} e^{-tz}[(zI - A)^{-1} - \frac{1}{z}I]u_0 dz$$

(3.11)

provided that $\alpha > 0$. After parametrizing the integral we get

$$u(t) = \int_{-\infty}^{\infty} F(t, \eta)d\eta$$

(3.12)

with

$$F(t, \eta) = -\frac{1}{2\pi i}(2a_I\eta - i)e^{-t(a_I\eta^2 + b_I - i\eta)}$$

$$\times \left\{[(a_I\eta^2 + b_I - i\eta)I - A]^{-1} - \frac{1}{a_I\eta^2 + b_I - i\eta}I\right\}u_0.$$

(3.13)

Following [62], we construct a quadrature rule for the integral in (2.34) by using the Sinc approximation on $(-\infty, \infty)$. For $1 \le p \le \infty$, we introduce the family $\mathbf{H}^p(D_d)$ of all vector-valued functions, which are analytic in the infinite strip D_d,

$$D_d = \{z \in \mathbb{C} : -\infty < \Re z < \infty, \ |\Im z| < d\}, \tag{3.14}$$

such that if $D_d(\epsilon)$ is defined for $0 < \epsilon < 1$ by

$$D_d(\epsilon) = \{z \in \mathbb{C} : |\Re z| < 1/\epsilon, \ |\Im z| < d(1 - \epsilon)\}, \tag{3.15}$$

then for each $\mathcal{F} \in \mathbf{H}^p(D_d)$ there holds $\|\mathcal{F}\|_{\mathbf{H}^p(D_d)} < \infty$ with

$$\|\mathcal{F}\|_{\mathbf{H}^p(D_d)} = \begin{cases} \lim_{\epsilon \to 0}(\int_{\partial D_d(\epsilon)} \|\mathcal{F}(z)\|^p |dz|)^{1/p} & \text{if } 1 \le p < \infty, \\ \lim_{\epsilon \to 0} \sup_{z \in \partial D_d(\epsilon)} \|\mathcal{F}(z)\| & \text{if } p = \infty. \end{cases} \tag{3.16}$$

Let

$$S(k,h)(x) = \frac{\sin\left[\pi(x - kh)/h\right]}{\pi(x - kh)/h} \tag{3.17}$$

be the k-th Sinc function with step size h, evaluated in x. Given $\mathcal{F} \in \mathbf{H}^p(D_d)$, $h > 0$ and positive integer N, let us use the notation

$$I(\mathcal{F}) = \int_{\mathbb{R}} \mathcal{F}(x)dx, \quad T_N(\mathcal{F},h) = h \sum_{k=-N}^{N} \mathcal{F}(kh),$$

$$T(\mathcal{F},h) = h \sum_{k=-\infty}^{\infty} \mathcal{F}(kh),$$

$$C(\mathcal{F},h) = \sum_{k=-\infty}^{\infty} \mathcal{F}(kh)S(k,h),$$

$$\eta_N(\mathcal{F},h) = I(\mathcal{F}) - T_N(\mathcal{F},h), \quad \eta(\mathcal{F},h) = I(\mathcal{F}) - T(\mathcal{F},h).$$

Applying the quadrature rule T_N with the vector-valued function (3.13) we obtain

$$u(t) = \exp(-tA)u_0 \approx u_N(t) = \exp_N(-tA)u_0 = h\left(\sum_{k=-N}^{N} F(kh,t)\right)u_0 \tag{3.18}$$

for integral (3.12).

Further, we show that this Sinc-quadrature approximation with a proper choice of h converges exponentially provided that the integrand can be analytically extended into a strip D_d. This property of the integrand depends on the choice of the integration path.

Taking into account (3.10) we get

$$\|F(t,\eta)\| \leq c\frac{e^{-t(a_I\eta^2 + b_I)}}{(1 + |\eta|)^{2\alpha}}\|A^\alpha u_0\|, \quad \forall t \geq 0, \ \alpha > 1/2 \tag{3.19}$$

(the condition $\alpha > 1/2$ guarantees the convergence of the integral (3.12)). The analysis of the integration parabola implies that the vector-valued function $F(\eta,t)$ can be analytically extended into the strip D_d and belongs to the class $\mathbf{H}^1(D_d)$ with respect to η, with the estimate

$$\|F(t,z)\|_{\mathbf{H}^1(D_d)} \leq c\frac{e^{-b_I t}}{2\alpha - 1}\|A^\alpha u_0\|, \quad \forall t \geq 0, \ \alpha > 1/2. \tag{3.20}$$

For our further analysis of the error $\eta_N(\mathcal{F},h) = \exp(-tA)u_0 - \exp_N(-tA)u_0$ of the quadrature rule (3.18) we use the following lemma from [30].

Lemma 3.1. *For any vector-valued function $f \in \mathbf{H}^1(D_d)$, there holds*

$$\eta(\tilde{f}, h) = \frac{i}{2} \int_{\mathbb{R}} \left\{ \frac{\tilde{f}(\xi - id^-)\mathrm{e}^{-\pi(d+i\xi)/h}}{\sin\left[\pi(\xi - id)/h\right]} - \frac{\tilde{f}(\xi + id^-)\mathrm{e}^{-\pi(d-i\xi)/h}}{\sin\left[\pi(\xi + id)/h\right]} \right\} d\xi, \quad (3.21)$$

which yields the estimate

$$\|\eta(\tilde{f}, h)\| \leq \frac{\mathrm{e}^{-\pi d/h}}{2\sinh(\pi d/h)} \|\tilde{f}\|_{\mathbf{H}^1(D_d)}. \quad (3.22)$$

If, in addition, $\tilde{f}$ satisfies on $\mathbb{R}$ the condition

$$\|\tilde{f}(x)\| \leq \frac{c\mathrm{e}^{-\beta x^2}}{(1 + x^2)^\sigma}, \quad 1/2 < \sigma \leq 1, \quad (3.23)$$

$$c, \beta > 0,$$

then

$$\|\eta_N(\tilde{f}, h)\| \leq \frac{2c}{2\sigma - 1} \left\{ 2\sigma \frac{exp(-\pi d/h)}{\sinh(\pi d/h)} + \frac{exp\left(-\beta(Nh)^2\right)}{(Nh)^{2\sigma - 1}} \right\}. \quad (3.24)$$

Taking into account the estimates (3.19), (3.20) and setting F for $\tilde{f}$, α for σ and ta_I for β we get the estimate

$$\|\eta_N(F, h)\| = \|\exp(-tA)u_0 - \exp_N(-tA)u_0\|$$
$$\leq c\frac{\mathrm{e}^{-b_I t}}{2\alpha - 1} \left\{ 2\alpha \frac{\exp(-\pi d/h)}{\sinh(\pi d/h)} + \frac{\exp\left(-a_I t(Nh)^2\right)}{(Nh)^{2\alpha - 1}} \right\} \|A^\alpha u_0\|. \quad (3.25)$$

Equalizing the exponents by setting $\pi d/h = a_I(Nh)^2$, we get the step-size of the quadrature

$$h = \sqrt[3]{\pi d/(a_I N^2)}. \quad (3.26)$$

Because $\sinh(\pi d/h) \geq \mathrm{e}^{\pi d/h}/2$, $\pi d/h = (\sqrt{a_I}\pi dN)^{2/3}$, $Nh = \sqrt[3]{\pi dN/a_I}$, $(Nh)^{2\alpha - 1} = (\pi dN/a_I)^{(2\alpha - 1)/3}$ and $d = b_0$ we get the following scale of estimates for the algorithm (3.18):

$$\|\eta_N(\mathcal{F}, h)\| = \|\exp(-tA)u_0 - \exp_N(-tA)u_0\| \quad (3.27)$$
$$\leq c\frac{\mathrm{e}^{-b_I t}}{2\alpha - 1} \left\{ 4\alpha \exp(-2(\sqrt{a_I}\pi b_0)^{2/3}N^{2/3}) + \frac{\exp\left(-a_I t(\pi b_0 N/a_I)^{2/3}\right)}{(\pi b_0 N/a_I)^{(2\alpha - 1)/3}} \right\} \|A^\alpha u_0\|,$$

with the step-size (3.26). Thus, we have proven the following statement.

Theorem 3.2. *Let A be a strongly P-positive operator in a Banach space X with the resolvent satisfying (2.36) and with the spectral parabola given by (2.37). Then for the Sinc approximation (3.18) we have the estimate (3.27), i.e.,*

$$\| \exp(-tA)u_0 - exp_N(-tA)u_0 \| = \begin{cases} \mathcal{O}(e^{-c_1 N^{2/3}}) & \text{if } t > 0, \\ \mathcal{O}(N^{(2\alpha-1)/3}) & \text{if } t = 0, \end{cases} \tag{3.28}$$

provided that $u_0 \in D(A^\alpha)$, $\alpha > 1/2$.

Remark 3.3. The above developed algorithm possesses the two sequential levels of parallelism: one can compute all $\hat{u}(z_p)$ at Step 2 and the solution $u(t) = e^{-At}u_0$ at different time values $(t_1, t_2, \ldots, t_M)$.

3.1.2 New algorithm for the operator exponential with an exponential convergence estimate including $t = 0$

We consider the following representation of the operator exponential:

$$e^{-At}u_0 = \frac{1}{2\pi i} \int_{\Gamma_I} e^{-zt}(zI - A)^{-1}u_0 dz. \tag{3.29}$$

Our aim is to approximate this integral by a quadrature with exponential convergence rate including $t = 0$. It is very important to keep in mind the representation of the solution of the inhomogeneous initial value problem (3.1) by

$$u(t) = e^{-At}u_0 + \int_0^t e^{-A(t-\xi)} \tilde{f}(\xi) d\xi, \tag{3.30}$$

where the argument of the operator exponential under the integral becomes zero for $\xi = t$. Taking into account (2.27) for $m = 0$ one can see that we can use the representation

$$e^{-At}u_0 = \frac{1}{2\pi i} \int_{\Gamma_I} e^{-zt} \left[(zI - A)^{-1} - \frac{1}{z}I \right] u_0 dz \tag{3.31}$$

instead of (3.29) (for $t > 0$ the integral from the second summand is equal to zero due to the analyticity of the integrand inside of the integration path) and this integral represents the solution of problem (3.1) for $u_0 \in D(A^\alpha)$, $\alpha > 0$. We call the hyperbola

$$\Gamma_0 = \{z(\xi) = a_0 \cosh \xi - ib_0 \sinh \xi : \xi \in (-\infty, \infty), \ b_0 = a_0 \tan \varphi\} \tag{3.32}$$

the spectral hyperbola, which has a path through the vertex $(a_0, 0)$ of the spectral angle and possesses asymptotes which are parallel to the rays of the spectral angle Σ (see Fig. 3.1). We choose as an integration path the hyperbola

$$\Gamma_I = \{z(\xi) = a_I \cosh \xi - ib_I \sinh \xi : \xi \in (-\infty, \infty)\}. \tag{3.33}$$

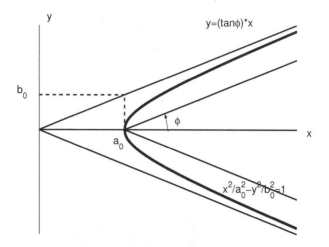

Figure 3.1: Spectral characteristics of the operator A.

The parametrization of the integral (3.31) by (3.33) implies

$$u(t) = \frac{1}{2\pi i} \int_{-\infty}^{\infty} \mathcal{F}(t,\xi)d\xi \qquad (3.34)$$

with

$$\mathcal{F}(t,\xi) = F_A(t,\xi)u_0,$$
$$F_A(t,\xi) = e^{-z(\xi)t}(a_I \sinh \xi - ib_I \cosh \xi) \left[(z(\xi)I - A)^{-1} - \frac{1}{z(\xi)}I \right]. \qquad (3.35)$$

To estimate $\|\mathcal{F}(t,\xi)\|$, we need an estimate for

$$|z'(\xi)/z(\xi)| = (a_I \sinh \xi - ib_I \cosh \xi)/(a_I \cosh \xi - ib_I \sinh \xi)$$
$$= \sqrt{(a_I^2 \tanh^2 \xi + b_I^2)/(b_I^2 \tanh^2 \xi + a_I^2)}.$$

The quotient under the square root takes its maximum at $v = 0$ as a function of $v = \tanh^2 \xi \in [0,1]$, since the sign of the first derivative coincides with the sign of $a_I^4 - b_I^4 = -a_0^4 \sin \varphi / \cos^4 \varphi$, i.e., we have

$$|z'(\xi)/z(\xi)| \leq b_I/a_I. \qquad (3.36)$$

Supposing $u_0 \in D(A^\alpha)$, $0 < \alpha < 1$, using (3.36) and Theorem 2.4, we can estimate the integrand on the real axis $\xi \in \mathbb{R}$ for each $t \geq 0$ by

$$
\begin{aligned}
\|\mathcal{F}(t,\xi)\| &\leq e^{-a_I t \cosh \xi} \frac{(1+M)K\sqrt{a_I^2 \sinh^2 \xi + b_I^2 \cosh^2 \xi}}{(a_I^2 \cosh^2 \xi + b_I^2 \sinh^2 \xi)^{(1+\alpha)/2}} \|A^\alpha u_0\| \\
&\leq (1+M)K \frac{b_I}{a_I} \frac{e^{-a_I t \cosh \xi}}{(a_I^2 \cosh^2 \xi + b_I^2 \sinh^2 \xi)^{\alpha/2}} \|A^\alpha u_0\| \\
&\leq (1+M)K \frac{b_I}{a_I} \left(\frac{2}{a_I}\right)^\alpha e^{-a_I t \cosh \xi - \alpha|\xi|} \|A^\alpha u_0\|, \quad \xi \in \mathbb{R},\ t \geq 0.
\end{aligned}
\tag{3.37}
$$

Let us show that the function $\mathcal{F}(t,\xi)$ can be analytically extended with respect to ξ into a strip of a width d_1. After changing ξ to $\xi + i\nu$, the integration hyperbola Γ_I is translated to the curve

$$
\begin{aligned}
\Gamma(\nu) &= \{z(w) = a_I \cosh(\xi + i\nu) - i b_I \sinh(\xi + i\nu) : \xi \in (-\infty, \infty)\} \\
&= \{z(w) = a(\nu)\cosh\xi - ib(\nu)\sinh\xi : \xi \in (-\infty, \infty)\}
\end{aligned}
\tag{3.38}
$$

with

$$
a(\nu) = a_I \cos\nu + b_I \sin\nu = \sqrt{a_I^2 + b_I^2}\sin(\nu + \phi/2),
$$

$$
b(\nu) = b_I \cos\nu - a_I \sin\nu = \sqrt{a_I^2 + b_I^2}\cos(\nu + \phi/2), \tag{3.39}
$$

$$
\cos\frac{\phi}{2} = \frac{b_I}{\sqrt{a_I^2 + b_I^2}}, \quad \sin\frac{\phi}{2} = \frac{a_I}{\sqrt{a_I^2 + b_I^2}}.
$$

The analyticity of the function $\mathcal{F}(t, \xi + i\nu)$, $|\nu| < d_1/2$ can be violated when the resolvent becomes unbounded. Thus, we must choose d_1 in such a way as to provide that the hyperbola $\Gamma(\nu)$ for $\nu \in (-d_1/2, d_1/2)$ remains in the right half-plane of the complex plane, and, in addition, when $\nu = -d_1/2$ $\Gamma(\nu)$ coincides with the imaginary axis, for $\nu = d_1/2$, $\Gamma(\nu)$ coincides with the spectral hyperbola, and for all $\nu \in (-d_1/2, d_1/2)$, $\Gamma(\nu)$ does not intersect the spectral sector. Then we choose the hyperbola $\Gamma(0)$ as the integration hyperbola.

This implies the following system of equations

$$
\begin{cases}
a_I \cos(d_1/2) + b_I \sin(d_1/2) = a_0, \\
b_I \cos(d_1/2) - a_I \sin(d_1/2) = a_0 \tan\varphi, \\
a_I \cos(-d_1/2) + b_I \sin(-d_1/2) = 0,
\end{cases}
\tag{3.40}
$$

from which we get

$$
\begin{cases}
2a_I \cos(d_1/2) = a_0, \\
b_I = a_0 \sin(d_1/2) + b_0 \cos(d_1/2), \\
a_I = a_0 \cos(d_1/2) - b_0 \sin(d_1/2).
\end{cases}
\tag{3.41}
$$

Eliminating a_I from the first and the third equations of (3.41) we get $a_0 \cos d_1 = b_0 \sin d_1$, i.e., $d_1 = \pi/2 - \varphi$ with $\cos\varphi = \frac{a_0}{\sqrt{a_0^2 + b_0^2}}$, $\sin\varphi = \frac{b_0}{\sqrt{a_0^2 + b_0^2}}$. Thus, if we choose the parameters of the integration hyperbola as follows:

$$
\begin{aligned}
a_I &= a_0 \cos\left(\frac{\pi}{4} - \frac{\varphi}{2}\right) - b_0 \sin\left(\frac{\pi}{4} - \frac{\varphi}{2}\right) \\
&= \sqrt{a_0^2 + b_0^2} \cos\left(\frac{\pi}{4} + \frac{\varphi}{2}\right) = a_0 \frac{\cos\left(\frac{\pi}{4} + \frac{\varphi}{2}\right)}{\cos\varphi}, \\
b_I &= a_0 \sin\left(\frac{\pi}{4} - \frac{\varphi}{2}\right) + b_0 \cos\left(\frac{\pi}{4} - \frac{\varphi}{2}\right) \\
&= \sqrt{a_0^2 + b_0^2} \sin\left(\frac{\pi}{4} + \frac{\varphi}{2}\right) = a_0 \frac{\sin\left(\frac{\pi}{4} + \frac{\varphi}{2}\right)}{\cos\varphi},
\end{aligned}
\tag{3.42}
$$

the vector-valued function $\mathcal{F}(t, w)$ is analytic with respect to $w = \xi + i\nu$ in the strip

$$
D_{d_1} = \{w = \xi + i\nu : \ \xi \in (-\infty, \infty), \ |\nu| < d_1/2\}, \tag{3.43}
$$

for all $t \geq 0$. Now, estimate (3.37) takes the form

$$
\begin{aligned}
\|\mathcal{F}(t, \xi)\| &\leq C(\varphi, \alpha) e^{-a_I t \cosh \xi - \alpha|\xi|} \|A^\alpha u_0\| \\
&\leq C(\varphi, \alpha) e^{-\alpha|\xi|} \|A^\alpha u_0\|, \ \xi \in \mathbb{R}, \ t \geq 0
\end{aligned}
\tag{3.44}
$$

with

$$
C(\varphi, \alpha) = (1 + M) K \tan\left(\frac{\pi}{4} + \frac{\varphi}{2}\right) \left(\frac{2 \cos\varphi}{a_0 \cos\left(\frac{\pi}{4} + \frac{\varphi}{2}\right)}\right)^\alpha. \tag{3.45}
$$

Comparing (3.42) with (3.39), we get $\phi = \pi/2 - \varphi$ and

$$
\begin{aligned}
a(\nu) &= a_I \cos\nu + b_I \sin\nu = \frac{a_0 \cos\left(\pi/4 + \varphi/2 - \nu\right)}{\cos\varphi}, \\
b(\nu) &= b_I \cos\nu - a_I \sin\nu = \frac{a_0 \sin\left(\pi/4 + \varphi/2 - \nu\right)}{\cos\varphi}, \\
0 &< a(\nu) < a_0, \quad a_0 \tan\varphi < b(\nu) < \frac{a_0}{\cos\varphi}.
\end{aligned}
\tag{3.46}
$$

Choosing $d = d_1 - \delta$ for an arbitrarily small positive δ and for $w \in D_d$ gets the estimate (compare with (3.37))

$$\|\mathcal{F}(t, w)\| \leq e^{-a(\nu)t \cosh \xi} \frac{(1 + M)K\sqrt{a^2(\nu)\sinh^2 \xi + b^2(\nu)\cosh^2 \xi}}{(a^2(\nu)\cosh^2 \xi + b^2(\nu)\sinh^2 \xi)^{(1+\alpha)/2}} \|A^\alpha u_0\|$$

$$\leq (1 + M)K\frac{b(\nu)}{a(\nu)} \frac{e^{-a(\nu)t \cosh \xi}}{(a^2(\nu)\cosh^2 \xi + b^2(\nu)\sinh^2 \xi)^{(\alpha/2)}} \|A^\alpha u_0\|$$

$$\leq (1 + M)K\frac{b(\nu)}{a(\nu)} \left(\frac{2}{a(\nu)}\right)^\alpha e^{-a(\nu)t \cosh \xi - \alpha|\xi|} \|A^\alpha u_0\|$$

$$\leq (1 + M)K \tan\left(\frac{\pi}{4} + \frac{\varphi}{2} - \nu\right) \left(\frac{2\cos\varphi}{a_0 \cos(\pi/4 + \varphi/2 - \nu)}\right)^\alpha$$

$$\times e^{-\alpha|\xi|} \|A^\alpha u_0\|, \quad \forall w \in D_d. \tag{3.47}$$

Accounting for the fact that the integrals over the vertical sides of the rectangle $D_d(\epsilon)$ vanish as $\epsilon \to 0$ the previous estimate implies

$$\|\mathcal{F}(t, \cdot)\|_{\mathbf{H}^1(D_d)} \leq \|A^\alpha u_0\|[C_-(\varphi, \alpha, \delta) + C_+(\varphi, \alpha, \delta)] \int_{-\infty}^{\infty} e^{-\alpha|\xi|} d\xi$$

$$= C(\varphi, \alpha, \delta)\|A^\alpha u_0\| \tag{3.48}$$

with

$$C(\varphi, \alpha, \delta) = \frac{2}{\alpha}[C_+(\varphi, \alpha, \delta) + C_-(\varphi, \alpha, \delta)],$$

$$C_\pm(\varphi, \alpha, \delta) = (1 + M)K \tan\left(\frac{\pi}{4} + \frac{\varphi}{2} \pm \frac{d}{2}\right) \left(\frac{2\cos\varphi}{a_0 \cos\left(\frac{\pi}{4} + \frac{\varphi}{2} \pm \frac{d}{2}\right)}\right)^\alpha. \tag{3.49}$$

Note that the constant $C(\varphi, \alpha, \delta)$ tends to ∞ as $\alpha \to 0$ or $\delta \to 0$, $\varphi \to \pi/2$.

We approximate integral (3.34) by the Sinc-quadrature

$$u_N(t) = \frac{h}{2\pi i} \sum_{k=-N}^{N} \mathcal{F}(t, z(kh)) \tag{3.50}$$

with the error estimate

$$\|\eta_N(\mathcal{F}, h)\| = \|u(t) - u_N(t)\|$$

$$\leq \left\|u(t) - \frac{h}{2\pi i}\sum_{k=-\infty}^{\infty} \mathcal{F}(t, z(kh))\right\| + \left\|\frac{h}{2\pi i}\sum_{|k|>N} \mathcal{F}(t, z(kh))\right\|$$

$$\leq \frac{1}{2\pi}\frac{e^{-\pi d/h}}{2\sinh(\pi d/h)}\|\mathcal{F}\|_{\mathbf{H}^1(D_d)} \tag{3.51}$$

$$+ \frac{C(\varphi, \alpha)h\|A^\alpha u_0\|}{2\pi} \sum_{k=N+1}^{\infty} \exp[-a_I t \cosh(kh) - \alpha kh]$$

$$\leq \frac{c\|A^\alpha u_0\|}{\alpha} \left\{\frac{e^{-\pi d/h}}{\sinh(\pi d/h)} + \exp[-a_I t \cosh((N+1)h) - \alpha(N+1)h]\right\}$$

where the constant c does not depend on h, N, t. Equalizing both exponentials for $t = 0$ by

$$\frac{2\pi d}{h} = \alpha(N+1)h \tag{3.52}$$

we get

$$h = \sqrt{\frac{2\pi d}{\alpha(N+1)}} \tag{3.53}$$

for the step-size. With this step-size, the error estimate

$$\|\eta_N(\mathcal{F}, h)\| \le \frac{c}{\alpha} \exp\left(-\sqrt{\frac{\pi d\alpha}{2}}(N+1)\right) \|A^\alpha u_0\| \tag{3.54}$$

holds true with a constant c independent of t, N. In the case $t > 0$ the first summand in the expression of $\exp[-a_I t \cosh((N+1)h) - \alpha(N+1)h]$ of (3.51) contributes mainly to the error order. By setting $h = c_1 \ln N/N$ with a positive constant c_1, we have an error

$$\|\eta_N(\mathcal{F}, h)\| \le c\left[e^{-\pi dN/(c_1 \ln N)} + e^{-c_1 a_I tN/2 - c_1 \alpha \ln N}\right] \|A^\alpha u_0\|, \tag{3.55}$$

for a fixed t, where c is a positive constant. Thus, we have proved the following result.

Theorem 3.4. *Let A be a densely defined strongly positive operator and $u_0 \in D(A^\alpha)$, $\alpha \in (0,1)$, then the Sinc-quadrature (3.50) represents an approximate solution of the homogeneous initial value problem (3.1) (i.e., $u(t) = e^{-At}u_0$) and possesses a uniform (with respect to $t \ge 0$) exponential convergence rate with estimate (3.51). This estimate is of the order $\mathcal{O}(e^{-c\sqrt{N}})$ uniformly in $t \ge 0$ for $h = \mathcal{O}(1/\sqrt{N})$ and of the order $\mathcal{O}\left(\max\left\{e^{-\pi dN/(c_1 \ln N)}, e^{-c_1 a_I tN/2 - c_1 \alpha \ln N}\right\}\right)$ for each fixed $t > 0$ provided that $h = c_1 \ln N/N$.*

Remark 3.5. Two other algorithms of the convergence order $\mathcal{O}(e^{-c\sqrt{N}})$ uniformly in $t \ge 0$ was proposed in [18, Remark 4.3 and (2.41)]. One of them used a sum of resolvents applied to u_0 provided that the operator coefficient is bounded. Another one was based on the representation

$$u(t) = \int_\Gamma z^{-\sigma} e^{-zt}(zI - A)^{-1} A^\sigma u_0 \tag{3.56}$$

valid for $u_0 \in D(A^\sigma), \sigma > 1$. Approximating the integral (after parametrizing Γ) by a Sinc-quadrature, one gets a short sum of resolvents applied to $A^\sigma u_0$ ([18, see (2.41)], [68]). The last vector must be computed first, where $\sigma = 2$ is the common choice. It is easy to see that, for $u_0 \in D(A^\sigma)$, both representations (3.56) and (3.31) are, in fact, equivalent, although the orders of computational stages (i.e., the algorithms) are different depending on the integral representation in use. But in the case $\sigma < 1$, the convergence theory for (3.56) was not presented in [18, 68]. Our representation (3.31) produces a new approximation through a short sum of modified resolvents $(zI-A)^{-1}-z^{-1}I$ applied to u_0 with the convergence properties given by Theorem 3.4. An approximation of the accuracy order $\mathcal{O}(e^{-cN/\ln N})$ for each fixed $t > 0$ to the operator exponential generated by a strongly P-positive operator and using a short sum of the usual resolvents was recently proposed in [19].

Remark 3.6. Note that taking $(zI - A)^{-1}$ in (3.31) instead of $(zI - A)^{-1} - \frac{1}{z}I$ we obtain a difference given by

$$D_I(t) = -\frac{1}{2\pi i} \int_{\Gamma_I} e^{-zt} \frac{1}{z} u_0 dz. \tag{3.57}$$

For the integration path Γ_I and $t = 0$, this difference can be calculated analytically. Actually, taking into account that the real part is an odd function and the integral of it in the sense of Cauchy is equal to zero, we further get

$$
\begin{aligned}
D_I(0) &= -\frac{1}{2\pi i} P.V. \int_{\Gamma_I} \frac{1}{z} u_0 dz = -\frac{1}{2\pi} \int_{-\infty}^{\infty} \frac{a_I b_I d\xi}{a_I^2 \cosh^2 \xi + b_I \sinh^2 \xi} u_0 \\
&= \frac{a_I b_I}{2\pi} \int_{-\infty}^{\infty} \frac{d(\tanh \xi)}{a_I^2 + b_I^2 \tanh^2 \xi} u_0 = \frac{1}{\pi} \arctan \frac{b_I}{a_I} u_0 = \frac{1}{\pi} \left(\frac{\pi}{4} + \frac{\varphi}{2} \right) u_0
\end{aligned}
\tag{3.58}
$$

for the integral of the imaginary part, where the factor in the front of u_0 is less than $1/2$. It means that one can expect a large error for sufficiently small t by using $(zI - A)^{-1}$ instead of $(zI - A)^{-1} - \frac{1}{z}I$ in (3.31). This point can be observed in the next example. Note that for $t > 0$, integral (3.57) is equal to 0 due to the analyticity of the integrand inside of the integration path.

Example 3.7. Let us choose $a_0 = \pi^2$, $\varphi = 0.8\pi/2$, then Table 3.1 gives the values of $\|D_I(t)\|/\|u_0\|$ for various t.

t	$\|D_I(t)\|/\|u_0\|$
0	0.45
$0.1 \cdot 10^{-8}$	0.404552
$0.1 \cdot 10^{-7}$	0.081008
$0.1 \cdot 10^{-6}$	0.000257
$0.1 \cdot 10^{-5}$	$0.147153 \cdot 10^{-6}$

Table 3.1: The unremovable error obtained by using the resolvent instead of $(zI - A)^{-1} - \frac{1}{z}I$.

3.1.3 Exponentially convergent algorithm II

Figure 3.2 shows the behavior of the integrand $\mathcal{F}(t, \xi)$ in (3.34) with the operator A defined by $D(A) = \{v(x) : v \in H^2(0, 1), v(0) = v(1) = 0\}$, $Au = -\frac{d^2 u}{dx^2}$.

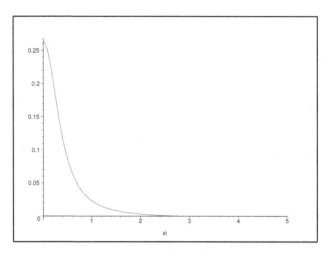

Figure 3.2: The behavior of the integrand $\mathcal{F}(t,\xi)$ in (3.34).

One can observe that the integrand is concentrated on a small finite interval and decays rapidly outside of this interval. This fact can be a cause for slow convergence of the previous algorithm for not large N. In this section we construct another exponentially convergent quadrature which accounts for the behavior of the integrand.

Due to the fact that the integrand exponentially decays on the infinite interval, it is reasonable to use an exponentially convergent quadrature rule on a finite interval, where the integrand is mostly concentrated, and to estimate the residual part. We represent integral (3.34) in the form

$$u(t) = \frac{1}{2\pi i} \int_{-\infty}^{\infty} \mathcal{F}(t,\xi)d\xi = I_1(t) + I_2(t) \qquad (3.59)$$

with

$$
\begin{aligned}
I_1(t) &= \frac{1}{2\pi i} \int_{-\beta}^{\beta} \mathcal{F}(t,\xi)d\xi, \\
I_2(t) &= \frac{1}{2\pi i} \int_{-\infty}^{-\beta} \mathcal{F}(t,\xi)d\xi + \frac{1}{2\pi i} \int_{\beta}^{\infty} \mathcal{F}(t,\xi)d\xi.
\end{aligned}
\qquad (3.60)
$$

Using estimate (3.45) gets

$$\left\| \frac{\|A^\alpha u_0\|}{2\pi i} \int_\beta^\infty F(t,\xi) d\xi \right\| \le \frac{\|A^\alpha u_0\|}{2\pi} (1+M) K \tan\left(\frac{\pi}{4} + \frac{\varphi}{2}\right)$$

$$\times \left(\frac{2}{\sqrt{a_0^2 + b_0^2}\cos\left(\frac{\pi}{4} + \frac{\varphi}{2}\right)} \right)^\alpha \int_\beta^\infty e^{-\sqrt{a_0^2 + b_0^2}\cos\left(\frac{\pi}{4} + \frac{\varphi}{2}\right)t \cosh\xi - \alpha|\xi|} d\xi \qquad (3.61)$$

$$\le C_1(\varphi, \alpha) \|A^\alpha u_0\| e^{-\sqrt{a_0^2 + b_0^2}\cos\left(\frac{\pi}{4} + \frac{\varphi}{2}\right)t\cosh\beta} \int_\beta^\infty e^{-\alpha|\xi|} d\xi$$

$$\le C_1(\varphi, \alpha) \|A^\alpha u_0\| e^{-\sqrt{a_0^2 + b_0^2}\cos\left(\frac{\pi}{4} + \frac{\varphi}{2}\right)t\cosh\beta} e^{-\alpha|\beta|}$$

with the constant

$$C_1(\varphi, \alpha) = \frac{(1+M)K}{2\pi\alpha} \tan\left(\frac{\pi}{4} + \frac{\varphi}{2}\right) \left(\frac{2}{\sqrt{a_0^2 + b_0^2}\cos\left(\frac{\pi}{4} + \frac{\varphi}{2}\right)} \right)^\alpha.$$

independent of β. This constant tends to ∞ if $\alpha \to 0$ or $\varphi \to \pi/2$. Analogously, one gets

$$\left\| \frac{1}{2\pi i} \int_{-\infty}^{-\beta} F(t,\xi) \right\| \le C_1(\varphi, \alpha) \|A^\alpha u_0\| e^{-\sqrt{a_0^2 + b_0^2}\cos\left(\frac{\pi}{4} + \frac{\varphi}{2}\right)t\cosh\beta} e^{-\alpha|\beta|}, \qquad (3.62)$$

which yields the estimate

$$\|I_2\| \le 2C_1(\varphi, \alpha) \|A^\alpha u_0\| e^{-\sqrt{a_0^2 + b_0^2}\cos\left(\frac{\pi}{4} + \frac{\varphi}{2}\right)t\cosh\beta} e^{-\alpha|\beta|}. \qquad (3.63)$$

Following [62] let us define the eye-shaped region (see Fig. 3.3)

$$\mathcal{D} = D_d^2 = \left\{ z \in \mathbb{C} : \left| \arg\left(\frac{z+\beta}{z-\beta} \right) \right| < d \right\} \qquad (3.64)$$

for $d \in (0, \pi)$ and the class $\mathbf{L}_{\kappa,\mu}(\mathcal{D})$ of all vector-valued functions holomorphic in $\mathcal{D}$ which additionally satisfy the inequality

$$\|F(z)\| \le c|z + \beta|^{\kappa-1}|z - \beta|^{\mu-1} \qquad (3.65)$$

with some positive real constants c, κ, μ.

In the previous section, we have shown that $\mathcal{F}(t,\xi)$ can be analytically extended into the strip D_d of width $2d$ which is symmetric with respect to the real axis. The equation of the boundary of the eye-shaped region in Cartesian coordinates is $\frac{2\beta y}{x^2 + y^2 - \beta^2} = \pm\tan d_1$. For $x = 0$, the maximal value of y, which remains in the analyticity region, is $y = d$ and we get the equation $\frac{2\beta d}{d^2 - \beta^2} = \pm\tan d_1$ for the maximal value of d_1 implying

$$d_1 \asymp d/\beta \qquad (3.66)$$

for sufficiently large β.

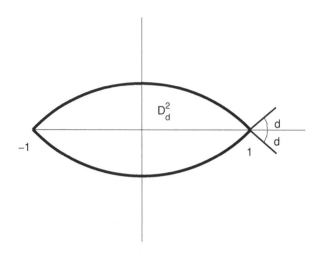

Figure 3.3: The eye-shaped region.

Given N and a function $F(\xi) \in \mathbf{L}_{\kappa,\mu}$, which can be analytically extended into an eye-shaped domain $D^2_{d_1}$, let us define (see [62])

$$\epsilon = \min{(\kappa, \mu)}, \quad \delta = \max{(\kappa, \mu)},$$

$$h = \left(\frac{2\pi d}{\epsilon n}\right)^{1/2},$$

$$M_l = \begin{cases} N & \text{if } \epsilon = \kappa, \\ [\mu N/\kappa] & \text{otherwise}, \end{cases} \qquad M_u = \begin{cases} [\kappa N/\mu] & \text{if } \epsilon = \kappa, \\ N & \text{otherwise}. \end{cases}$$

(3.67)

Then

$$\left\| \int_{-\beta}^{\beta} F(\xi)d\xi - 2\beta h \sum_{-M_l}^{M_u} \frac{e^{kh}}{(1+e^{kh})^2} F(z_k) \right\| \le c e^{-\sqrt{2\pi d_1 \epsilon N}},$$

(3.68)

where the nodes are $z_k = \frac{-\beta + \beta e^{kh}}{1 + e^{kh}}$.

Using this quadrature and taking into account that $\mathcal{F}(t, \xi) \in \mathbf{L}_{1,1}(\mathcal{D})$ (with respect to ξ) we get the following Sinc-quadrature approximation for I_1:

$$I_1(t) \approx I_{1,N}(t) = \frac{2\beta h}{2\pi i} \sum_{-N}^{N} \frac{e^{kh}}{(1+e^{kh})^2} \mathcal{F}(t, z_k),$$

$$h = \left(\frac{2\pi d_1}{N}\right)^{1/2}.$$

(3.69)

with the approximation error

$$\|\eta_{N,1}(t)\| \le c\|A^\alpha u_0\| e^{-\sqrt{2\pi d_1 N}}.$$

(3.70)

Setting

$$u(t) = e^{-At}u_0 \approx I_1(t) \tag{3.71}$$

we obtain the full approximation error

$$\|u(t) - \mathring{I}_{1,N}\| = \|e^{-At}u_0 - I_{1,N}\| \leq \|\eta_{N,1}\| + \|I_2(t)\|$$
$$\leq c\|A^\alpha u_0\|(e^{-\sqrt{2\pi d_1 N}} + e^{-\sqrt{a_0^2 + b_0^2}\cos\left(\frac{\pi}{4} + \frac{\varphi}{2}\right)t}\cosh\beta\, e^{-\alpha|\beta|}). \tag{3.72}$$

Equating the exponents and taking into account (3.66) we get that $h = \left(\frac{2\pi d}{N^{4/3}}\right)^{1/2}$ and

$$\|e^{-At}u_0 - I_{1,N}(t)\| \leq c\|A^\alpha u_0\|e^{-c_1 N^{1/3}} \tag{3.73}$$

provided

$$\beta \asymp N^{1/3} \tag{3.74}$$

Example 3.8. We consider problem (3.1) with $u_0 = (1 - x)x^2$ and the operator A defined by $D(A) = \{v(x) : v \in H^2(0,1), v(0) = v(1) = 0\}$, $Au = -\frac{d^2 u}{dx^2}$. It is easy to see that $u_0 \in D(A^1)$ and the exact solution is given by

$$u(t,x) = -\frac{4}{\pi^3}\sum_1^\infty \frac{(2(-1)^k + 1)}{k^3}e^{-\pi^2 k^2 t}\sin(\pi k x).$$

One can show that

$$(zI - A)^{-1}u_0 - u_0/z = \frac{1}{z}(zI - A)^{-1}Au_0 \tag{3.75}$$
$$= \frac{6x - 2}{z^2} - \frac{\cos\left[\sqrt{z}(1/2 - x)\right]}{z^2\cos(\sqrt{z}/2)} + 3\frac{\sin\left[\sqrt{z}(1/2 - x)\right]}{z^2\sin(\sqrt{z}/2)}.$$

Table 3.2 gives the solution computed by the algorithm (3.50) with $h = \sqrt{2\pi/N}$ (the first column) and by algorithm (3.71) with $h = \sqrt{2\pi/N^{4/3}}$ (the second column). The exact solution is $u(0, 1/2) = u_0(1/2) = 1/8$. This example shows that, even though algorithm (3.50) is better for sufficiently large N, the algorithm (3.71) can be better for relatively small N. Besides, the table confirms the exponential convergence of both algorithms.

3.1.4 Inhomogeneous differential equation

In this section, we consider the solution of the inhomogeneous problem (3.1)

$$u(t) = u_h(t) + u_p(t), \tag{3.76}$$

where

$$u_h(t) = e^{-At}u_0, \quad u_p(t) = \int_0^t e^{-A(t-s)}f(s)ds. \tag{3.77}$$

N	A1	A2
8	0.147319516168	0.121686777535
16	0.131006555144	0.124073586590
32	0.125894658654	0.124809057018
64	0.125055464496	0.124952849785
128	0.125000975782	0.124995882473
256	0.125000002862	0.124999802171

Table 3.2: The solution for $t = 0$, $x = 1/2$ by the algorithms (3.50) (A1) and
(3.71) (A2).

Note, that an algorithm for convolution integrals like the ones from previous
sections was described (without theoretical justification) in [62] based on the Sinc
quadratures. If the Laplace transform of $f(t)$ is known and is sectorial, then one
can use an algorithm based on inversion of the Laplace transform of convolution
[48]. In the case when the Laplace transform of $f(t)$ is not known we propose in
this section a discretization different from [62] and [48].

Using representation (3.31) of the operator exponential gives

$$
\begin{aligned}
u_p(t) &= \int_0^t \frac{1}{2\pi i} \int_{\Gamma_I} e^{-z(t-s)}[(zI - A)^{-1} - \frac{1}{z}I]f(s)dzds \\
&= \frac{1}{2\pi i} \int_{\Gamma_I} \left[(z(\xi)I - A)^{-1} - \frac{1}{z(\xi)}I \right] \int_0^t e^{-z(\xi)(t-s)}f(s)dsz'(\xi)d\xi, \quad (3.78)
\end{aligned}
$$

$$z(\xi) = a_I \cosh \xi - ib_I \sinh \xi.$$

Replacing the first integral by quadrature (3.50) we get

$$
u_p(t) \approx u_{ap}(t) = \frac{h}{2\pi i} \sum_{k=-N}^{N} z'(kh) \left[(z(kh)I - A)^{-1} - \frac{1}{z(kh)}I \right] f_k(t) \quad (3.79)
$$

with

$$
f_k(t) = \int_0^t e^{-z(kh)(t-s)}f(s)ds, \quad k = -N, \ldots, N. \quad (3.80)
$$

To construct an exponentially convergent quadrature for these integrals we change
the variables by

$$
s = \frac{t}{2}(1 + \tanh \xi) \quad (3.81)
$$

and get

$$
f_k(t) = \int_{-\infty}^{\infty} \mathcal{F}_k(t, \xi)d\xi, \quad (3.82)
$$

instead of (3.80), where

$$
\mathcal{F}_k(t, \xi) = \frac{t}{2 \cosh^2 \xi} \exp[-z(kh)t(1 - \tanh \xi)/2]f(t(1 + \tanh \xi)/2). \quad (3.83)
$$

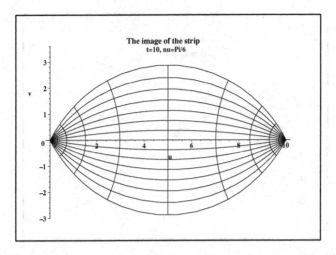

Figure 3.4: The image of the strip for $t = 10$, $\nu = \pi/6$.

Note that equation (3.81) with the complex variables $z = \xi + i\nu$ and $w = u + iv$ represents the conformal mapping $w = \psi(z) = t[1 + \tanh z]/2$, $z = \phi(w) = \frac{1}{2}\ln\frac{t-w}{w}$ of the strip D_ν onto the domain $\mathcal{A}_\nu$ (compare with the domain D_ν^2 in [62]). The integrand can be estimated on the real axis as

$$\|\mathcal{F}_k(t, \xi)\| \leq \frac{t}{2\cosh^2\xi}\exp[-a_I\cosh(kh)t(1 - \tanh\xi)/2] \tag{3.84}$$
$$\times \|f(t(1 + \tanh\xi)/2)\| \leq 2te^{-2|\xi|}\|f(t(1 + \tanh\xi)/2)\|.$$

Lemma 3.9. Let the right-hand side $f(t)$ in (3.1) for $t \in [0, \infty]$ be analytically extended into the sector $\Sigma_f = \{\rho e^{i\theta_1} : \rho \in [0, \infty], |\theta_1| < \varphi\}$ and for all complex $w \in \Sigma_f$ we have

$$\|f(w)\| \leq ce^{-\delta|\Re w|} \tag{3.85}$$

with $\delta \in (0, \sqrt{2}a_0]$; then the integrand $\mathcal{F}_k(t, \xi)$ can be analytically extended into the strip D_{d_1}, $0 < d_1 < \varphi/2$ and it belongs to the class $H^1(D_{d_1})$ with respect to ξ, where a_0, φ are the spectral characterizations (2.23) of A.

Proof. Let us investigate the domain in the complex plane for which the function $\mathcal{F}(t, \xi)$ can be analytically extended to the real axis $\xi \in \mathbb{R}$. Replacing ξ with $\xi + i\nu$, $\xi \in (-\infty, \infty)$, $|\nu| < d_1$, in the integrand we get

$$\tanh(\xi + i\nu) = \frac{\sinh\xi\cos\nu + i\cosh\xi\sin\nu}{\cosh\xi\cos\nu + i\sinh\xi\sin\nu} = \frac{\sinh(2\xi) + i\sin(2\nu)}{2(\cosh^2\xi - \sin^2\nu)}, \tag{3.86}$$
$$1 \pm \tanh(\xi + i\nu) = q_r^\pm + iq_i^\pm$$

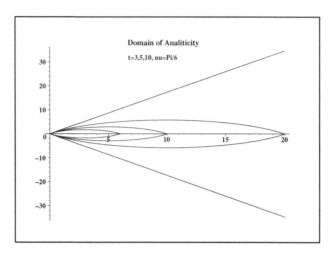

Figure 3.5: The domains of the analyticity of the integrand for $t = 3, 5, 10$, $\nu = \pi/6$.

in particular for the argument of f, where

$$
q_r^{\pm}(\xi, \nu) = 1 \pm \frac{\sinh 2\xi}{2(\cosh^2 \xi - \sin^2 \nu)} = \frac{e^{\pm 2\xi} + \cos(2\nu)}{2(\cosh^2 \xi - \sin^2 \nu)},
$$

$$
q_i^{\pm}(\xi, \nu) = \pm \frac{\sin 2\nu}{2(\cosh^2 \xi - \sin^2 \nu)}.
$$

$$(3.87)$$

The denominator in (3.86) is not equal to zero for all $\xi \in (-\infty, \infty)$ provided that $\nu \in (-\pi/2, \pi/2)$. It is easy to see that we have

$$
0 \le q_r^{\pm}(\xi, \nu) \le 2,
$$

$$
|q_i^{\pm}(\xi, \nu)| \le |\tan \nu|,
$$

$$(3.88)$$

for $\xi \in (-\infty, \infty)$, i.e., for each fixed t, ν and for $\xi \in (-\infty, \infty)$, the parametric curve $\Gamma_A(t)$ from (3.86) given by (in the coordinates μ, η)

$$
\mu = \frac{t}{2} q_r^-(\xi, \nu),
$$

$$
\eta = \frac{t}{2} q_i^-(\xi, \nu)
$$

$$(3.89)$$

is closed and forms the angle with the real axis at the origin,

$$
\theta = \theta(\nu) = \arctan|\lim_{\xi \to \infty} q_i^-(\xi, \nu)/q_r^-(\xi, \nu)| = \arctan(\tan(2\nu)) = 2\nu. \quad (3.90)
$$

For $\nu \in (-\pi/4, \pi/4)$ the domain $\mathcal{A}(t)$ inside of $\Gamma_A(t)$ lies in the right half-plane and for $t \to \infty$ fills the sector $\Sigma_f(\nu) = \{z = \rho e^{i\psi} : \rho \in (0, \infty), \ \psi \in (-\nu, \nu), \ \nu \in (0, \pi/4)\}$ (see Fig. 3.5). Taking into account (3.42), we have

$$\|\mathcal{F}(t,\xi+i\nu)\| \leq \frac{t}{2(\cosh^2\xi-\sin^2\nu)}$$

$$\times |\exp\{-\frac{t[a_I\cosh(kh)-ib_I\sinh(kh)]}{2}[q_r^+ + iq_i^+]\}|$$

$$\times \|f(t(1+\tanh(\xi+i\nu))/2)\|$$

$$\leq \frac{t}{2(\cosh^2\xi-\sin^2\nu)} \tag{3.91}$$

$$\times \exp\{-\frac{ta_0[\cosh(kh)\cos(\pi/4+\varphi/2)(\cos(2\nu)+e^{2\xi})]}{2(\cosh^2\xi-\sin^2\nu)}$$

$$-\frac{ta_0[\sinh(kh)\cos(\pi/4+\varphi/2)\sin(2\nu)]}{2(\cosh^2\xi-\sin^2\nu)}\}$$

$$\times \|f(t(1+\tanh(\xi+i\nu))/2)\|$$

(note that $\nu \in (-\pi/2, \pi/2)$ provides that $\cosh^2\xi - \sin^2\nu > 0$ for all $\xi \in (-\infty, \infty)$).
Because we assume that

$$\|f(w)\| \leq ce^{-\delta|\Re w|}, \quad \delta > 0, \tag{3.92}$$

omitting the second summand in the argument of the exponential and replacing $\cosh(kh)$ by 1 gets the inequality

$$\|\mathcal{F}(t,\xi+i\nu)\| \leq \frac{ct}{2(\cosh^2\xi-\sin^2\nu)}\exp\left\{\frac{t[-\Delta e^{2\xi}-\delta e^{-2\xi}/2]}{2(\cosh^2\xi-\sin^2\nu)}\right\} \tag{3.93}$$

where

$$\frac{a_0}{2} \leq \Delta = a_0\frac{\cos(\varphi/2+\pi/4)}{\cos\varphi} = \frac{a_0}{\sqrt{2}\sqrt{1+\sin\varphi}} \leq \frac{a_0}{\sqrt{2}}. \tag{3.94}$$

Due to assumption $\delta \leq \sqrt{2}a_0$, we have $\delta/2 \leq \Delta$ and the last estimate yields

$$\|\mathcal{F}(t,\xi+i\nu)\| \leq \frac{ct}{2(\cosh^2\xi-\sin^2\nu)}\exp\{-\frac{t\delta\cosh(2\xi)}{2(\cosh^2\xi-\sin^2\nu)}\}. \tag{3.95}$$

Denoting $w = t\Delta\cosh(2\xi)/[2(\cosh^2\xi-\sin^2\nu)]$ and using (3.94) and the inequality $we^{-w} \leq e^{-1}\ \forall\ w \geq 0$ gives

$$\int_{-\infty}^{\infty}\|\mathcal{F}(t,\xi+i\nu)\|d\xi$$

$$\leq \int_{-\infty}^{\infty}\frac{ct}{2(\cosh^2\xi-\sin^2\nu)}\exp\{-\frac{t\delta\cosh(2\xi)}{2(\cosh^2\xi-\sin^2\nu)}\}d\xi$$

$$= \int_{-\infty}^{\infty}\frac{1}{\Delta\cosh(2\xi)}we^{-w}d\xi \leq \frac{c}{e\Delta}\int_{-\infty}^{\infty}\frac{1}{\cosh(2\xi)}d\xi \tag{3.96}$$

$$\leq \frac{2c}{e\Delta}\int_{-\infty}^{\infty}e^{-2|\xi|}d\xi = \frac{2c}{e\Delta} \leq \frac{4c}{a_0 e}.$$

This estimate yields $\mathcal{F}_k(t,\xi) \in H^1(D_{d_1})$ with respect to ξ. $\qquad\qquad\square$

The assumptions of Lemma 3.9 can be weakened when we consider problem (3.1) on a finite interval $(0, T]$.

Lemma 3.10. *Let the right-hand side $f(t)$ in (3.1) for $t \in [0, T]$ be analytically extended into the domain $\mathcal{A}(T)$, then the integrand $\mathcal{F}_k(t, \xi)$ can be analytically extended into the strip D_{d_1}, $0 < d_1 < \varphi/2$ and belongs to the class $H^1(D_{d_1})$ with respect to ξ.*

Proof. The proof is analogous to that of Lemma 3.9 but with constants depending on T.   □

Let the assumptions of Lemma 3.9 hold. Then we can use the following quadrature rule to compute the integrals (3.82) (see [62], p. 144):

$$f_k(t) \approx f_{k,N}(t) = h \sum_{p=-N}^{N} \mu_{k,p}(t) f(\omega_p(t)), \tag{3.97}$$

where

$$\mu_{k,p}(t) = \frac{t}{2} \exp\{-\frac{t}{2} z(kh)[1 - \tanh{(ph)}]\} / \cosh^2{(ph)},$$
$$\omega_p(t) = \frac{t}{2}[1 + \tanh{(ph)}], \quad h = \mathcal{O}(1/\sqrt{N}), \tag{3.98}$$
$$z(\xi) = a_I \cosh \xi - i b_I \sinh \xi.$$

Substituting (3.97) into (3.79), we get the following algorithm to compute an approximation $u_{ap,N}(t)$ to $u_{ap}(t)$:

$$u_{ap,N}(t) = \frac{h}{2\pi i} \sum_{k=-N}^{N} z'(kh)[(z(kh)I - A)^{-1} - \frac{1}{z(kh)} I]$$
$$\times h \sum_{p=-N}^{N} \mu_{k,p}(t) f(\omega_p(t)). \tag{3.99}$$

The next theorem characterizes the error of this algorithm.

Theorem 3.11. *Let A be a densely defined strongly positive operator with the spectral characterization a_0, φ and the right-hand side $f(t) \in D(A^\alpha)$, $\alpha > 0$ (for $t \in [0, \infty]$) be analytically extended into the sector $\Sigma_f = \{\rho e^{i\theta_1} : \rho \in [0, \infty], |\theta_1| < \varphi\}$ where the estimate*

$$\|A^\alpha f(w)\| \leq c_\alpha e^{-\delta_\alpha |\Re w|}, \quad w \in \Sigma_f \tag{3.100}$$

with $\delta_\alpha \in (0, \sqrt{2}a_0]$ holds. Then algorithm (3.99) converges with the error estimate

$$\|\mathcal{E}_N(t)\| = \|u_p(t) - u_{ap,N}(t)\| \leq c e^{-c_1 \sqrt{N}} \tag{3.101}$$

uniformly in t with positive constants c, c_1 depending on α, φ, a_0 and independent of N.

Proof. Let us set

$$R_k(t) = f_k(t) - f_{k,N}(t). \tag{3.102}$$

We get for the error

$$\mathcal{E}_N(t) = u_p(t) - u_{ap,N}(t) = r_{1,N}(t) + r_{2,N}(t), \tag{3.103}$$

where

$$\begin{aligned} r_{1,N}(t) &= u_p(t) - u_{ap}(t), \\ r_{2,N}(t) &= u_{ap}(t) - u_{ap,N}(t). \end{aligned} \tag{3.104}$$

Using estimate (3.54) (see also Theorem 3.4), we get the estimate

$$\begin{aligned} &\|r_{1,N}(t)\| \\ &= \|\int_0^t \{\frac{1}{2\pi i}\int_{-\infty}^\infty F_A(t-s,\xi)d\xi - \frac{h}{2\pi i}\sum_{k=-N}^N F_A(t-s,kh)\}f(s)ds\| \\ &\leq \frac{c}{\alpha}\exp\left(-\sqrt{\frac{\pi d\alpha}{2}}(N+1)\right)\int_0^t \|A^\alpha f(s)\|ds, \end{aligned} \tag{3.105}$$

for $r_{1,N}(t)$, where $F_A(t,\xi)$ is the operator defined in (3.35). Due to (2.31), the error $r_{2,N}(t)$ takes the form

$$\begin{aligned} \|r_{2,N}(t)\| &= \|\frac{h}{2\pi i}\sum_{k=-N}^N z'(kh)[(z(kh)I - A)^{-1} - \frac{1}{z(kh)}I]R_k(t)\| \\ &\leq \frac{h(1+M)K}{2\pi}\sum_{k=-N}^N \frac{|z'(kh)|}{|z(kh)|^{1+\alpha}}\|A^\alpha R_k(t)\|. \end{aligned} \tag{3.106}$$

The estimate (3.84) yields

$$\|A^\alpha F(t,\xi)\| \leq 2te^{-2|\xi|}\|A^\alpha f(\frac{t}{2}(1+\tanh\xi))\|. \tag{3.107}$$

Due to Lemma 3.9 the assumption $\|A^\alpha f(w)\| \leq c_\alpha e^{-\delta_\alpha|\Re w|} \ \forall w \in \Sigma_f$ guarantees that $A^\alpha f(w) \in H^1(D_{d_1})$ and $A^\alpha F_k(t,w) \in H^1(D_{d_1})$. Then we are in a situation analogous to that of Theorem 3.2.1, p. 144 from [62] with $A^\alpha f(w)$ instead of f.

This implies

$$\|A^\alpha R_k(t)\| = \|A^\alpha(f_k(t) - f_{k,N}(t))\|$$

$$= \|\int_{-\infty}^{\infty} A^\alpha \mathcal{F}_k(t,\xi)d\xi - h\sum_{k=-\infty}^{\infty} A^\alpha \mathcal{F}_k(t,kh)\| + \|h\sum_{|k|>N} A^\alpha \mathcal{F}_k(t,kh)\|$$

$$\leq \frac{e^{-\pi d_1/h}}{2\sinh(\pi d_1/h)}\|\mathcal{F}_k(t,w)\|_{H^1(D_{d_1})}$$

$$+ h\sum_{|k|>N} 2te^{-2|kh|}\|A^\alpha f(\frac{t}{2}(1+\tanh kh))\| \tag{3.108}$$

$$\leq ce^{-2\pi d_1/h}\|A^\alpha f(t,w)\|_{H^1(D_{d_1})}$$

$$+ h\sum_{|k|>N} 2te^{-2|kh|}c_\alpha \exp\{-\delta_\alpha\frac{t}{2}(1-\tanh kh)\}$$

$$\leq ce^{-c_1\sqrt{N}},$$

where positive constants $c_\alpha, \delta_\alpha, c, c_1$ do not depend on t, N, k. Now, (3.106) takes the form

$$\|r_{2,N}(t)\| = \frac{h}{2\pi i}\sum_{k=-N}^{N} z'(kh)[(z(kh)I - A)^{-1} - \frac{1}{z(kh)}I]R_k(t) \tag{3.109}$$

$$\leq ce^{-c_1\sqrt{N}}S_N$$

with

$$S_N = \sum_{k=-N}^{N} h\frac{|z'(kh)|}{|z(kh)|^{1+\alpha}}.$$

Using the estimate (3.36) and

$$|z(kh)| = \sqrt{a_I^2\cosh^2(kh) + b_I^2\sinh^2(kh)}$$

$$\geq a_I\cosh(kh) \geq a_I e^{|kh|}/2, \tag{3.110}$$

the last sum can be estimated by

$$|S_N| \leq \frac{c}{\sqrt{N}}\sum_{k=-N}^{N} e^{-\alpha|k|/\sqrt{N}} \leq c\int_{-\sqrt{N}}^{\sqrt{N}} e^{-\alpha t}dt \leq c/\alpha. \tag{3.111}$$

Taking into account (3.108), (3.111) we get from (3.109)

$$\|r_{2,N}(t)\| \leq ce^{-c_1\sqrt{N}}. \tag{3.112}$$

The assertion of the theorem follows now from (3.103), (3.105).   □

Example 3.12. We consider the inhomogeneous problem (3.1) with the operator A defined by

$$D(A) = \{u(x) \in H^2(0,1) : u(0) = u(1) = 0\},$$
$$Au = -u''(x) \ \forall u \in D(A). \tag{3.113}$$

The initial function is $u_0 = u(0,x) = 0$ and the right-hand side $f(t)$ is given by

$$f(t,x) = x^3(1-x)^3 \frac{1-t^2}{(1+t^2)^2} - \frac{6t}{1+t^2} x(1-x)(5x^2 - 5x + 1). \tag{3.114}$$

It is easy to see that the exact solution is $u(t,x) = x^3(1-x)^3 \frac{t}{1+t^2}$. The algorithm (3.99) was implemented for $t = 1$, $x = 1/2$ in Maple 8 with Digits=16. Table 3.3 shows an exponential decay of the error $\varepsilon_N = |u(1,1/2) - u_{ap,N}(1)|$ with growing N.

N	ε_N
8	0.485604499
16	0.184497471
32	0.332658314 e-1
64	0.196729786 e-2
128	0.236757688 e-4
256	0.298766899 e-7

Table 3.3: The error of algorithm (3.99) for $t = 0$, $x = 1/2$.

3.2 Algorithms without accuracy saturation for first-order evolution equations in Hilbert and Banach spaces

3.2.1 Introduction

We consider the evolution problem

$$\frac{du}{dt} + A(t)u = f(t), \quad t \in (0,T]; \quad u(0) = u_0, \tag{3.115}$$

where $A(t)$ is a densely defined closed (unbounded) operator with the domain $D(A)$ independent of t in a Banach space X, u_0, u_{01} are given vectors and $f(t)$ is a given vector-valued function. We suppose the operator $A(t)$ to be strongly positive; i.e., there exists a positive constant M_R independent of t such that, on the rays and outside a sector $\Sigma_\theta = \{z \in \mathbb{C} : 0 \leq \arg(z) \leq \theta, \theta \in (0, \pi/2)\}$, the following resolvent estimate holds:

$$\|(zI - A(t))^{-1}\| \leq \frac{M_R}{1 + |z|}. \tag{3.116}$$

This assumption implies that there exists a positive constant c_κ such that (see [14], p. 103)

$$\|A^\kappa(t)e^{-sA(t)}\| \leq c_\kappa s^{-\kappa}, \quad s > 0, \quad \kappa \geq 0. \tag{3.117}$$

Our further assumption is that there exists a real positive ω such that

$$\|e^{-sA(t)}\| \leq e^{-\omega s} \quad \forall s, \, t \in [0, T] \tag{3.118}$$

(see [54], Corollary 3.8, p. 12, for corresponding assumptions on $A(t)$). Let us also assume that the following conditions hold true:

$$\|[A(t) - A(s)]A^{-\gamma}(t)\| \leq \tilde{L}_{1,\gamma}|t - s| \quad \forall t, \, s, \, 0 \leq \gamma \leq 1, \tag{3.119}$$

$$\|A^\beta(t)A^{-\beta}(s) - I\| \leq \tilde{L}_\beta|t - s| \quad \forall t, \, s \in [0, T]. \tag{3.120}$$

In addition, we can suppose that

$$A(t) = \sum_{k=0}^{m_A} A_k t^k, \; f(t) = \sum_{k=0}^{m_f} f_k t^k. \tag{3.121}$$

Note that efficient approximations without accuracy saturation or with exponential accuracy for the solution operator (3.115) (with an unbounded operator A independent of t) were proposed in [15, 16, 20, 23, 27, 68]. Some of them were considered in section 3.1.

The aim of this section is to get an algorithm without accuracy saturation and an exponentially convergent algorithm for the solution of the problem (3.115). We use a piecewise constant approximation of the operator $A(t)$ and an exact integral corollary of these equations on the Chebyshev grid which is approximated by the collocation method. The operator exponential (for equation (3.115)) with stationary operator involved in the algorithm can be computed by the Sinc approximations from section 3.1 (see also [16, 23, 27]).

We begin with an example which shows the practical relevance for the assumptions above.

Example 3.13. Let $q(t) \geq q_0 > 0$, $t \in [0, T]$, be a given function from the Hölder class with the exponent $\alpha \in (0, 1]$. We consider the operator $A(t)$ defined by

$$D(A(t)) = \{u(x) \in H^4(0, 1) : u(0) = u''(0) = u(1) = u''(1) = 0\},$$

$$A(t)u = \left[\frac{d^2}{dx^2} - q(t)\right]^2 u = \frac{d^4 u}{dx^4} - 2q(t)\frac{d^2 u}{dx^2} + q^2(t)u \quad \forall u \in D(A(t)) \tag{3.122}$$

with domain independent of t. It is easy to show that

$$D(A^{1/2}(t)) = \{u(x) \in H^2(0,1) : u(0) = u(1) = 0\},$$

$$A^{1/2}(t) = -\frac{d^2 u}{dx^2} + q(t)u \quad \forall u \in D\left(A^{1/2}(t)\right),$$

$$A^{-1/2}(t) = \int_0^1 G(x,\xi;t)v(\xi)\,d\xi,$$

(3.123)

where the Green function is given by

$$G(x,\xi;t) = \frac{1}{\sqrt{q(t)}\sinh\sqrt{q(t)}} \begin{cases} \sinh\left(\sqrt{q(t)}x\right)\sinh\left(\sqrt{q(t)}(1-\xi)\right), & \text{if } x \le \xi, \\ \sinh\left(\sqrt{q(t)}\xi\right)\sinh\left(\sqrt{q(t)}(1-x)\right), & \text{if } \xi \le x. \end{cases}$$

(3.124)

Then we have the relation

$$[A(t) - A(s)]A^{-1/2}(t)v$$

$$= [q(t) - q(s)]\left\{-2\frac{d^2}{dx^2} + [q(t) + q(s)]\right\}\int_0^1 G(x,\xi;t)v(\xi)\,d\xi$$

(3.125)

$$= [q(t) - q(s)]\left\{2v(x) - [q(t) - q(s)]\int_0^1 G(x,\xi;t)v(\xi)\,d\xi\right\},$$

which leads to the estimate

$$\|[A(t) - A(s)]A^{-1/2}(t)v(\cdot)\|_{C[0,1]\to C[0,1]}$$

$$\le L|t - s|^\alpha \left\{2\|v\|_{C[0,1]} + L|t - s|^\alpha \frac{1}{2\sqrt{q(t)}}\tanh\left(\sqrt{q(t)}/2\right)\|v\|_{C[0,1]}\right\},$$

(3.126)

where L is the Hölder constant. This inequality yields

$$\|[A(t) - A(s)]A^{-1/2}(t)\|_{C[0,1]\to C[0,1]}$$

$$\le L\left\{2 + LT^\alpha \tanh\left(\sqrt{q(t)}/2\right)/(2\sqrt{q(t)})\right\}|t - s|^\alpha,$$

(3.127)

i.e., condition (3.119) is fulfilled with $\gamma = 1/2$ provided that $\alpha = 1$. Let us prove the condition (3.120). We have

$$[A^{1/2}(t)A^{-1/2}(s) - I]v = \left[-\frac{d^2}{dx^2} + q(t)\right]\int_0^1 G(x,\xi;s)v(\xi)\,d\xi - v(x)$$

(3.128)

$$= [q(t) - q(s)]\int_0^1 G(x,\xi;s)v(\xi)\,d\xi,$$

from which it follows that

$$\|A^{1/2}(t)A^{-1/2}(s) - I\|_{C[0,1]\to C[0,1]} \le L\frac{\tanh\left(\sqrt{q(t)}/2\right)}{2\sqrt{q(t)}}|t - s|^\alpha;$$

(3.129)

i.e., condition (3.120) is fulfilled with $\beta = 1/2, \delta = \alpha = 1$.

Remark 3.14. It is clear that, in general, inequalities (3.119) and (3.120) hold true for elliptic operators with $\gamma = 1$, $\beta = 1$.

Remark 3.15. Assumption (3.121) is not restrictive due to stability results from [29]. The two initial value problems

$$\frac{du}{dt} + A(t)u = f(t), \quad u(0) = u_0 \tag{3.130}$$

and

$$\frac{dv}{dt} + B(t)v = g(t), \quad v(0) = v_0 \tag{3.131}$$

with densely defined, closed operators $A(t)$, $B(t)$ having a common domain $D\big(A(t)\big) = D\big(B(t)\big)$ independent of t were considered. The following assumptions were made.

1. There exist bounded inverse operators $A^{-1}(t)$, $B^{-1}(t)$ and for the resolvents, $R_{A(t)}(z) = (z - A(t))^{-1}$, $R_{B(t)}(z) = (z - B(t))^{-1}$ we have

$$\|R_{A(t)}(z)\| \le \frac{1}{1 + |z|}, \; \|R_{B(t)}(z)\| \le \frac{1}{1 + |z|} \quad (\theta + \epsilon \le |\arg z| \le \pi) \tag{3.132}$$

   for all $\theta \in (0, \pi/2)$, $\epsilon > 0$ uniformly in $t \in [0, T]$.

2. Operators $A(t), B(t)$ are strongly differentiable on $D\big(A(t)\big) = D\big(B(t)\big)$.

3. There exists a constant M such that

$$\big\|A^\beta(s)B^{-\beta}(s)\big\| \le M. \tag{3.133}$$

4. For the evolution operators $U_A(t, s), U_B(t, s)$ we have

$$\|A(t)U_A(t, s)\| \le \frac{C}{t - s}, \quad \|B(t)U_B(t, s)\| \le \frac{C}{t - s}. \tag{3.134}$$

5. There exist positive constants C, C_β such that

$$\big\|A^\rho(t)A^{-\rho}(s) - I\big\| \le C|t - s|^\alpha \tag{3.135}$$

   and

$$\|A^\beta(t)U_A(t, s)\| \le \frac{C_\beta}{|t - s|^\beta}, \; \|B^\beta(t)U_B(t, s)\| \le \frac{C_\beta}{|t - s|^\beta} \tag{3.136}$$

   for $0 \le \beta < \alpha + \beta$.

The following stability result for Banach spaces was proved in [29] under these assumptions:

$$
\begin{aligned}
\|A^\beta(t)z(t)\| &= \|A^\beta(t)(u(t) - v(t))\| \\
&\leq M\|A^\beta(0)z(0)\| + c_\beta M \max_{0\leq s\leq T} \left\|[B(s) - A(s)]A^{-\beta}(s)\right\| \\
&\quad \times \frac{t^{1-\beta}}{1-\beta}\left\{\|B^\beta(0)v(0)\| + \int_0^t \|B^\beta(s)g(s)\|\,ds\right\} \\
&\quad + M\int_0^t \|B^\beta(s)g(s)\|\,ds.
\end{aligned}
\tag{3.137}
$$

It is possible to avoid the restriction $\beta < 1$ if we consider equations (3.130), (3.131) in a Hilbert space. In this case we assume that there exists an operator $C = C^* \geq c_0 I$ such that

$$
\left\|[A(s) - B(s)]C^{-1}\right\| \leq \delta,
\tag{3.138}
$$

and

$$
\begin{aligned}
(A(s)y, Cy) &\geq c_0\|Cy\|^2, \\
(B(s)y, Cy) &\geq c_0\|Cy\|^2 \quad \forall s \in [0, T], \;\; c_0 > 0.
\end{aligned}
\tag{3.139}
$$

Then the following stability estimate is fulfilled [29]:

$$
\begin{aligned}
\frac{1}{2}(Cz(t), z(t)) &+ (c_0 - \epsilon - \epsilon_1)\int_0^t \|Cz(s)\|^2 ds \\
&\leq \max_{0\leq s\leq T} \left\|[A(s) - B(s)]C^{-1}\right\|^2 \frac{(c_0 - \epsilon_2)^{-1}}{2\epsilon}\left[\frac{1}{4\epsilon_2}\int_0^t \|g(s)\|^2 ds + \frac{1}{2}(Cv_0, v_0)\right] \\
&\quad + \frac{1}{2\epsilon_1}\int_0^t \|f(s) - g(s)\|^2 ds + \frac{1}{2}(C(u_0 - v_0), u_0 - v_0),
\end{aligned}
\tag{3.140}
$$

with arbitrary positive numbers $\epsilon, \epsilon_1, \epsilon_2$ such that $\epsilon + \epsilon_1 < c_0, \epsilon_2 < c_0$ which stand for the stability with respect to the right-hand side, the initial condition and the coefficient stability. Note that an analogous estimate in the case of a finite dimensional Hilbert spaces and of a constant operator A was proved in [57, p. 62].

Example 3.16. Let $\Omega \subset \mathbb{R}^2$ be a polygon and let

$$
\mathcal{L}(x, t, D) = -\sum_{i,j=1}^2 \frac{\partial}{\partial x_i} a_{i,j}(x, t)\frac{\partial}{\partial x_j} + \sum_{j=1}^2 b_j(x, t)\frac{\partial}{\partial x_j} + c(x, t)
\tag{3.141}
$$

be a second-order elliptic operator with time-dependent real smooth coefficients satisfying the uniform ellipticity condition

$$
\sum_{i,j=1}^2 a_{ij}(x, t)\xi_i\xi_j \geq \delta_1|\xi|^2 \quad (\xi = (\xi_1, \xi_2) \in \mathbb{R})
\tag{3.142}
$$

with a positive constant δ_1. Taking $X = L^2(\Omega)$ and $V = H_0^1(\Omega)$ or $V = H^1(\Omega)$ according to the boundary condition

$$u = 0 \quad \text{on} \quad \partial\Omega \times (0, T) \tag{3.143}$$

or

$$\frac{\partial u}{\partial \nu_{\mathcal{L}}} + \sigma u = 0 \quad \text{on} \quad \partial\Omega \times (0, T), \tag{3.144}$$

we set

$$\mathcal{A}_t(u, v) = \sum_{i,j=1}^{2} \int_{\Omega} a_{i,j}(x, t) \frac{\partial u}{\partial x_i} \frac{\partial v}{\partial x_j} dx + \sum_{j=1}^{2} \int_{\Omega} b_j(x, t) \frac{\partial u}{\partial x_j} v \, dx$$
$$+ \int_{\Omega} c(x, t) uv \, dx + \int_{\partial\Omega} \sigma(x, t) uv \, dS \tag{3.145}$$

for $u, v \in V$. An m-sectorial operator $A(t)$ in X can be defined through the relation

$$\mathcal{A}_t(u, v) = \big(A(t)u, v\big), \tag{3.146}$$

where $u \in D(A(t)) \subset V$ and $v \in V$. The relation

$$D(A(t)) = H^2(\Omega) \cap H_0^1(\Omega) \tag{3.147}$$

follows for $V = H_0^1(\Omega)$ and

$$D\big(A(t)\big) = \left\{ v \in H^2(\Omega) \mid \frac{\partial v}{\partial \nu_{\mathcal{L}}} \text{ on } \partial\Omega \right\} \tag{3.148}$$

for $V = H^1(\Omega)$, when $\partial\Omega$ is smooth, for instance.

It was proven in [14, pp. 95–101] that all the assumptions above hold for such an operator $A(t)$.

As we will see below, the parameter γ from (3.119) plays an essential role for the construction and the analysis of discrete approximations and algorithms for problem (3.115).

3.2.2   Discrete first-order problem in the case $\gamma < 1$

For the sake of simplicity we consider problem (3.115) on the interval $[-1, 1]$ (if it is not the case, one can reduce problem (3.115) to this interval by the variable transform $t = 2t'/T - 1$, $t \in [-1, 1]$, $t' \in [0, T]$). We choose a mesh ω_n of n various points $\omega_n = \{t_k = \cos\frac{(2k-1)\pi}{2n}, \ k = 1, \dots, n\}$ on $[-1, 1]$ and set $\tau_k = t_k - t_{k-1}$,

$$\overline{A}(t) = A_k = A(t_k), \quad t \in (t_{k-1}, t_k], \tag{3.149}$$

where t_k are zeros of a Chebyshev orthogonal polynomial of the first kind $T_n(t) = \cos{(n \arccos t)}$. Let $t_\nu = \cos\theta_\nu$, $0 < \theta_\nu < \pi$, $\nu = 1, 2, \dots, n$, be zeros of the

Chebyshev orthogonal polynomial $T_n(t)$ taken in decreasing order. Then it is well known that (see [63], Ch. 6, Th. 6.11.12, [64], p. 123)

$$t_{\nu+1} - t_\nu < \tfrac{\pi}{n}, \quad \nu = 1, \ldots, n,$$
$$\tau_{\max} = \max_{1 \le k \le n} \tau_k < \tfrac{\pi}{n}. \tag{3.150}$$

Let us rewrite problem (3.115) in the form

$$\frac{du}{dt} + \overline{A}(t)u = [\overline{A}(t) - A(t)]u(t) + f(t),$$
$$u(0) = u_0 \tag{3.151}$$

from which we deduce

$$u(t) = e^{-A_k(t-t_{k-1})}u(t_{k-1})$$
$$+ \int_{t_{k-1}}^{t} e^{-A_k(t-\eta)}\{[A_k - A(\eta)]u(\eta) + f(\eta)\}d\eta, \ t \in [t_{k-1}, t_k]. \tag{3.152}$$

Since A_{k-1} and $e^{-A_{k-1}\tau_k}$ commute, assumption (3.120) yields

$$\|A_k^\beta A_p^{-\beta}\| \le 1 + \|A_k^\beta A_p^{-\beta} - I\| \le 1 + \tilde{L}_\beta |t_k - t_p| \le 1 + \tilde{L}_\beta T. \tag{3.153}$$

Let

$$P_{n-1}(t; u) = P_{n-1}u = \sum_{p=1}^{n} u(t_p)L_{p,n-1}(t) \tag{3.154}$$

be the interpolation polynomial for the function $u(t)$ on the mesh ω_n, let $y = (y_1, \ldots, y_n)$, $y_i \in X$ be a given vector, and let

$$P_{n-1}(t; y) = P_{n-1}y = \sum_{p=1}^{n} y_p L_{p,n-1}(t) \tag{3.155}$$

be the polynomial that interpolates y, where $L_{p,n-1} = \frac{T_n(t)}{T_n'(t_p)(t-t_p)}$, $p = 1, \ldots, n$, are the Lagrange fundamental polynomials.

Substituting $P_n(\eta; y)$ for $u(\eta)$ and y_k for $u(t_k)$ in (3.152), we arrive at the following system of linear equations with respect to the unknowns y_k:

$$y_k = e^{-A_k\tau_k}y_{k-1} + \sum_{p=1}^{n} \alpha_{kp}y_p + \phi_k, \quad k = 1, \ldots, n, \tag{3.156}$$

where

$$\alpha_{kp} = \int_{t_{k-1}}^{t_k} e^{-A_k(t_k-\eta)}[A_k - A(\eta)]L_{p,n-1}(\eta)d\eta,$$
$$\phi_k = \int_{t_{k-1}}^{t_k} e^{-A_k(t_k-\eta)}f(\eta)d\eta. \tag{3.157}$$

Remark 3.17. To compute α_{kp} and ϕ_k in an efficient way, we replace $A(t), f(t)$ by their interpolation polynomials (it is possible due to stability results (3.137), (3.140); see also [29]) and calculate the integrals analytically. We have

$$A(t) = \sum_{l=1}^{n} \frac{A_l}{t - t_l} \frac{T_n(t)}{T'_n(t_l)},$$

$$f(t) = \sum_{l=1}^{n} \frac{f_l}{t - t_l} \frac{T_n(t)}{T'_n(t_l)}, \quad f_l = f(t_l), \tag{3.158}$$

so that

$$\alpha_{kp} = \frac{1}{T'_n(t_p)} \sum_{l=1}^{n} \frac{1}{T'_n(t_l)} \int_{t_{k-1}}^{t_k} e^{-A_k(t_k - \eta)} \frac{T_n^2(\eta)}{(\eta - t_l)(\eta - t_p)} d\eta [A_k - A_l],$$

$$\phi_k = \sum_{l=1}^{n} \frac{f_k}{T'_n(t_l)} \int_{t_{k-1}}^{t_k} e^{-A_k(t_k - \eta)} \frac{T_n(\eta)}{\eta - t_l} d\eta. \tag{3.159}$$

Using the relation $2T_n^2(\eta) = 1 + 2T_{2n}(\eta)$, the polynomial $p_{2n-2}^{(l,p)} = \frac{T_n^2(\eta)}{(\eta - t_l)(\eta - t_p)}$ can be represented as (see [6])

$$p_{2n-2}^{(l,p)} = \frac{2T_{2n}(\eta) + 1}{2(\eta - t_l)(\eta - t_p)}$$

$$= \frac{1}{2(\eta - t_l)(\eta - t_p)} \left[2n \sum_{m=0}^{n} \frac{(-1)^m (2n - m - 1)!}{m!(2n - 2m)!} (2\eta)^{2n-2m} + 1 \right] \tag{3.160}$$

$$= \sum_{i=0}^{2n-2} q_i(l, p) \eta^{2n-2-i},$$

where the coefficients $q_i(l, p)$ can be calculated, for example, by the Horner scheme. Given $q_i(l, p)$, we furthermore find that

$$\alpha_{kp} = \frac{1}{T'_n(t_p)} \sum_{l=1}^{n} \frac{1}{T'_n(t_l)} \sum_{i=0}^{2n-2} q_i(l, p) I_{k,i} [A_k - A_l], \tag{3.161}$$

where

$$I_{k,i} = \int_{t_{k-1}}^{t_k} e^{-A_k(t_k - \eta)} \eta^{2n-2-i} d\eta$$

$$= \sum_{s=0}^{2n-2-i} (-1)^s (2n - 2 - i)(2n - 3 - i) \cdots$$

$$\times (2n - 2 - i - s + 1) A_k^{-s-1} t_k^{2n-2-i-s} \tag{3.162}$$

$$- \sum_{s=0}^{2n-2-i} (-1)^s (2n-2-i)(2n-3-i) \cdots$$
$$\times (2n-2-i-s+1) A_k^{-s-1} t_{k-1}^{2n-2-i-s} e^{-A_k \tau_k}.$$

Analogously one can also calculate ϕ_k.

For the error $z = (z_1, \dots, z_n)$, $z_k = u(t_k) - y_k$ we have the relations

$$z_k = e^{-A_k \tau_k} z_{k-1} + \sum_{p=1}^{n} \alpha_{kp} z_p + \psi_k, \quad k = 1, \dots, n, \tag{3.163}$$

where

$$\psi_k = \int_{t_{k-1}}^{t_k} e^{-A_k(t_k-\eta)} [A_k - A(\eta)][u(\eta) - P_n(\eta; u)] d\eta. \tag{3.164}$$

We introduce the matrix $S = \{s_{i,k}\}_{i,k=1}^n$:

$$S = \begin{pmatrix} I & 0 & 0 & \cdots & & 0 & 0 \\ -e^{-A_1\tau_1} & I & 0 & \cdots & & 0 & 0 \\ 0 & -e^{-A_2\tau_2} & I & \cdots & & 0 & 0 \\ \cdot & & & \cdots & & \cdot & \cdot \\ 0 & 0 & 0 & \cdots & & -e^{-A_{n-1}\tau_{n-1}} & I \end{pmatrix}, \tag{3.165}$$

the matrix $C = \{\tilde{\alpha}_{k,p}\}_{k,p=1}^n$ with $\tilde{\alpha}_{k,p} = A_k^\gamma \alpha_{k,p} A_p^{-\gamma}$ and the vectors

$$y = \begin{pmatrix} A_1^\gamma y_1 \\ \cdot \\ \cdot \\ A_n^\gamma y_n \end{pmatrix}, \quad f = \begin{pmatrix} A_1^\gamma \phi_1 \\ \cdot \\ \cdot \\ A_n^\gamma \phi_n \end{pmatrix}, \quad \tilde{f} = \begin{pmatrix} A_1^\gamma e^{-A_1\tau_1} u_0 \\ 0 \\ \cdot \\ \cdot \\ 0 \end{pmatrix}, \quad \psi = \begin{pmatrix} A_1^\gamma \psi_1 \\ \cdot \\ \cdot \\ A_n^\gamma \psi_n \end{pmatrix}. \tag{3.166}$$

It is easy to see that for

$$S^{-1} = \{s_{i,k}^{-1}\}_{i,k=1}^n$$

$$= \begin{pmatrix} I & 0 & \cdots & 0 & 0 \\ e^{-A_1\tau_1} & I & \cdots & 0 & 0 \\ e^{-A_2\tau_2} e^{-A_1\tau_1} & e^{-A_2\tau_2} & \cdots & 0 & 0 \\ \cdot & & \cdots & & \cdot \\ e^{-A_{n-1}\tau_{n-1}} \cdots e^{-A_1\tau_1} & e^{-A_{n-1}\tau_{n-1}} \cdots e^{-A_2\tau_2} & \cdots & e^{-A_{n-1}\tau_{n-1}} & I \end{pmatrix} \tag{3.167}$$

we have

$$S^{-1} S = \begin{pmatrix} I & 0 & \cdots & 0 \\ 0 & I & \cdots & 0 \\ \cdot & \cdot & \cdots & \cdot \\ 0 & 0 & \cdots & I \end{pmatrix}. \tag{3.168}$$

Remark 3.18. Using results of [16, 23], one can get a parallel and sparse approximation with an exponential convergence rate of operator exponentials in S^{-1} and, as a consequence, a parallel and sparse approximation of S^{-1}.

We get from (3.156), (3.163)

$$
\begin{aligned}
A_k^\gamma y_k &= \mathrm{e}^{-A_k \tau_k} A_k^\gamma y_{k-1} + \sum_{p=1}^n \tilde\alpha_{kp} A_p^\gamma y_p + A_k^\gamma \phi_k, \\
A_k^\gamma z_k &= \mathrm{e}^{-A_k \tau_k} A_k^\gamma z_{k-1} + \sum_{p=1}^n \tilde\alpha_{kp} A_p^\gamma z_p + A_k^\gamma \psi_k, \ k = 1, \dots, n,
\end{aligned}
\tag{3.169}
$$

or, in the matrix form,

$$
\begin{aligned}
Sy &= Cy + f - \tilde f, \\
Sz &= Cz + \psi
\end{aligned}
\tag{3.170}
$$

with

$$
z = \begin{pmatrix} A_1^\gamma z_1 \\ \cdot \\ \cdot \\ \cdot \\ A_n^\gamma z_n \end{pmatrix}.
\tag{3.171}
$$

Next, for a vector $v = (v_1, v_2, \dots, v_n)^T$ and a block operator matrix $A = \{a_{ij}\}_{i,j=1}^n$ we introduce the vector norm

$$
|||v||| \equiv |||v|||_\infty = \max_{1 \le k \le n} \|v_k\|
\tag{3.172}
$$

and the consistent matrix norm

$$
|||A||| \equiv |||A|||_\infty = \max_{1 \le i \le n} \sum_{j=1}^n \|a_{i,j}\|.
\tag{3.173}
$$

Due to (3.118) we get

$$
|||S^{-1}||| \le n.
\tag{3.174}
$$

In our forthcoming analysis we need the following auxiliary result.

Lemma 3.19. *The estimates*

$$
|||C||| \le c(1 + \tilde L_\gamma T) n^{\gamma-2} \ln n,
\tag{3.175}
$$

$$
|||S^{-1}C||| \le c(1 + \tilde L_\gamma T) n^{\gamma-1} \ln n
\tag{3.176}
$$

with a positive constant c independent of n hold true.

Proof. Assumption (3.119) together with (3.153) implies

$$
\begin{aligned}
\|\tilde{\alpha}_{kp}\| &= \|A_k^\gamma \alpha_{kp} A_p^{-\gamma}\| \\
&= \Big\| \int_{t_{k-1}}^{t_k} A_k^\gamma e^{-A_k(t_k - \eta)} [A_k - A(\eta)] A_p^{-\gamma} L_{p,n-1}(\eta) d\eta \Big\| \\
&\le \left(1 + \tilde{L}_\gamma T\right) \tau_{\max}^{1-\gamma} \int_{t_{k-1}}^{t_k} |L_{p,n-1}(\eta)| d\eta, \quad T = 2.
\end{aligned}
\tag{3.177}
$$

Using the well-known estimate for the Lebesgue constant Λ_n related to the Chebyshev interpolation nodes (see, e.g., [63, 64])

$$
\Lambda_n = \max_{\eta \in [-1,1]} \sum_{p=1}^{n} |L_{p,n-1}(\eta)| \le c \ln n
\tag{3.178}
$$

and (3.177), we have

$$
\begin{aligned}
\||C\|| &\le \max_{1 \le k \le n} \sum_{p=1}^{n} \|\tilde{\alpha}_{kp}\| \le (1 + \tilde{L}_\gamma T) \tau_{\max}^{2-\gamma} \max_{\eta \in [-1,1]} \sum_{p=1}^{n} |L_{p,n-1}(\eta)| \\
&\le (1 + \tilde{L}_\gamma T) \tau_{\max}^{2-\gamma} \Lambda_n \le c(1 + \tilde{L}_\gamma T) \tau_{\max}^{2-\gamma} \ln n \le c(1 + \tilde{L}_\gamma T) n^{\gamma-2} \ln n
\end{aligned}
\tag{3.179}
$$

with an appropriate positive constant c independent of n. Together with (3.174), this estimate implies

$$
\||S^{-1}C\|| \le c(1 + \tilde{L}_\gamma T) n^{\gamma-1} \ln n \to 0
\tag{3.180}
$$

as $n \to \infty$ provided that $\gamma < 1$. $\square$

Remark 3.20. We have reduced the interval length to $T = 2$ but we write T explicitly to underline the dependence of involved constants on T, in the general case.

Let Π_{n-1} be the set of all polynomials in t with vector coefficients of degree less than or equal to $n - 1$. Then the Lebesgue inequality

$$
\|u(\eta) - P_{n-1}(\eta; u)\|_{C[-1,1]} \equiv \max_{\eta \in [-1,1]} \|u(\eta) - P_{n-1}(\eta; u)\| \le (1 + \Lambda_n) E_n(u)
\tag{3.181}
$$

can be proved for vector-valued functions in complete analogy with [5, 63, 64] with the error of the best approximation of u by polynomials of degree not greater than $(n - 1)$,

$$
E_n(u) = \inf_{p \in \Pi_{n-1}} \max_{\eta \in [-1,1]} \|u(\eta) - p(\eta)\|.
\tag{3.182}
$$

Now, we can go over to the main result of this section.

Theorem 3.21. *Let assumptions* (3.116)–(3.120) *with* $\gamma < 1$ *hold. Then there exists a positive constant* c *such that the following hold.*

1. *For n large enough it holds that*

$$|||z||| \equiv |||y - u||| \leq cn^{\gamma - 1} \ln n E_n(A_0^\gamma u), \tag{3.183}$$

   where u is the solution of (3.115).

2. *The system of linear algebraic equations*

$$Sy = Cy + f \tag{3.184}$$

   with respect to the approximate solution y can be solved by the fixed-point iteration

$$y^{(k+1)} = S^{-1}Cy^{(k)} + S^{-1}(f + \tilde{f}), \ k = 0, 1, \ldots; \ y^{(0)} \text{ arbitrary}, \tag{3.185}$$

   with the convergence rate of a geometrical progression with the denominator $q \leq cn^{\gamma-1}\ln n < 1$ for n large enough.

Proof. From the second equation in (3.170), we get

$$z = S^{-1}Cz + S^{-1}\psi \tag{3.186}$$

from which, due to Lemma 3.19 and (3.174)

$$|||z||| \leq cn|||\psi||| \tag{3.187}$$

for n large enough. The last norm can be estimated in the following way:

$$
\begin{aligned}
|||\psi||| &= \max_{1 \leq k \leq n} \| \int_{t_{k-1}}^{t_k} \left[A_k^\gamma e^{-A_k(t_k - \eta)} \left[A_k - A(\eta) \right] A_k^{-\gamma} (A_k^\gamma A_0^{-\gamma})(A_0^\gamma u(\eta) \right. \\
&\qquad\qquad \left. - P_n(\eta; A_0^\gamma u)) \right] d\eta \| \\
&\leq (1 + \tilde{L}_\gamma T) \max_{1 \leq k \leq n} \| \int_{t_{k-1}}^{t_k} |t_k - \eta|^{-\gamma} |t_k - \eta| \|A_0^\gamma u(\eta) - P_n(\eta; A_0^\gamma u)\| d\eta \\
&\leq (1 + \tilde{L}_\gamma T)\tau_{\max}^{2-\gamma} \|A_0^\gamma u(\cdot) - P_{n-1}(\cdot; A_0^\gamma u)\|_{C[-1,1]} \\
&\leq (1 + \tilde{L}_\gamma T)\tau_{\max}^{2-\gamma}(1 + \Lambda_n)E_n(A_0^\gamma u) \leq cn^{\gamma-2}\ln n E_n(A_0^\gamma u)
\end{aligned}
\tag{3.188}
$$

and taking into account (3.187), we get the statement of the theorem.   □

3.2.3 Discrete first-order problem in the case $\gamma \leq 1$

In this section, we construct a new discrete approximation of problem (3.115) which is a little more complicated than approximation (3.156) of the previous section but possesses a higher convergence order and allows the case $\gamma = 1$.

Applying transform (3.152) (i.e., substituting $u(t)$ recursively), we get

$$u(t) = \left\{ e^{-A_k(t-t_{k-1})} + \int_{t_{k-1}}^{t} e^{-A_k(t-\eta)} \left[A_k - A(\eta)\right] e^{-A_k(\eta-t_{k-1})} d\eta \right\} u(t_{k-1})$$

$$+ \int_{t_{k-1}}^{t} e^{-A_k(t-\eta)} [A_k - A(\eta)] \int_{t_{k-1}}^{\eta} e^{-A_k(\eta-s)} [A_k - A(s)] u(s) ds d\eta$$

$$+ \int_{t_{k-1}}^{t} e^{-A_k(t-\eta)} \left\{ [A_k - A(\eta)] \int_{t_{k-1}}^{\eta} e^{-A_k(\eta-s)} f(s) ds + f(\eta) \right\} d\eta.$$

$$(3.189)$$

Setting $t = t_k$, we arrive at the relation

$$u(t_k) = S_{k,k-1} u(t_{k-1}) \qquad (3.190)$$

$$+ \int_{t_{k-1}}^{t_k} e^{-A_k(t_k-\eta)} [A_k - A(\eta)] \int_{t_{k-1}}^{\eta} e^{-A_k(\eta-s)} [A_k - A(s)] u(s) ds d\eta + \phi_k,$$

where

$$S_{k,k-1} = e^{-A_k \tau_k} + \int_{t_{k-1}}^{t_k} e^{-A_k(t_k-\eta)} [A_k - A(\eta)] e^{-A_k(\eta-t_{k-1})} d\eta,$$

$$\phi_k = \int_{t_{k-1}}^{t_k} e^{-A_k(t_k-\eta)} f(\eta) d\eta \qquad (3.191)$$

$$+ \int_{t_{k-1}}^{t_k} e^{-A_k(t_k-\eta)} [A_k - A(\eta)] \int_{t_{k-1}}^{\eta} e^{-A_k(\eta-s)} f(s) ds d\eta.$$

Substituting the interpolation polynomial $P_{n-1}(\eta; y)$ from the previous section for $u(\eta)$ and y_k for $u(t_k)$ in (3.190), we arrive at the following system of linear equations with respect to the unknowns y_k:

$$y_k = S_{k,k-1} y_{k-1} + \sum_{p=1}^{n} \alpha_{kp} y_p + \phi_k, \quad k = 1, \dots, n, \qquad (3.192)$$

where

$$\alpha_{kp} = \int_{t_{k-1}}^{t_k} e^{-A_k(t_k-\eta)} [A_k - A(\eta)] \int_{t_{k-1}}^{\eta} e^{-A_k(\eta-s)} [A_k - A(s)] L_{p,n-1}(s) ds d\eta.$$

$$(3.193)$$

Remark 3.22. Due to stability results (3.137), (3.140) (see also [29]) one can approximate the initial problems with polynomials $\tilde{A}(t)$, $\tilde{f}(t)$, for example, as interpolation polynomials for $A(t)$, $f(t)$.

With the aim of getting a computational algorithm for α_{kp}, we write down formula (3.193) in the form

$$\alpha_{kp} = \int_{t_{k-1}}^{t_k} \int_{t_{k-1}}^{\eta} e^{-A_k(t_k-\eta)}[A_k - A(\eta)]e^{-A_k(\eta-s)}d\eta[A_k - A(s)]L_{p,n-1}(s)ds. \tag{3.194}$$

In order to calculate the inner integral, we represent

$$A(\eta) = A_k + (t_k - \eta)B_{1,k} + \cdots + (t_k - \eta)^{n-1}B_{n-1,k}. \tag{3.195}$$

Then we have to calculate integrals of the type

$$\tilde{\alpha}_{kp} = \int_{t_{k-1}}^{\eta} e^{-A_k(t_k-\eta)}(t_k - \eta)^p B_{p,k}e^{-A_k(\eta-s)}d\eta. \tag{3.196}$$

Analogously to [29], using the representation by the Dunford-Cauchy integrals and the residue theorem under assumption of the strong P-positiveness [15, 23, 16] of the operator $A(t)$, one can get

$$\tilde{\alpha}_{kp} = \frac{p!}{2\pi i} \int_{\Gamma_I} e^{-z(t_k-\eta)}(A_k - zI)^{-p-1}B_{p,k}(zI - A_k)^{-1}dz, \tag{3.197}$$

where Γ_I is an integration parabola enveloping the spectral parabola of the strongly P-positive operator $A(t)$. Now, using (3.195), (3.197) formula (3.194) can be written down as

$$\alpha_{kp} = -\frac{1}{2\pi i} \int_{\Gamma_I} \sum_{p=1}^{n-1} p!(A_k - zI)^{-p-1}B_{p,k}(zI - A_k)^{-1}$$

$$\times \int_{t_{k-1}}^{t_k} e^{-z(t_k-s)}[A_k - A(s)]L_{p,n-1}(s)dsdz. \tag{3.198}$$

The inner integral in this formula can be calculated analogously as in (3.161) and the integral along Γ_I can be calculated explicitly using the residue theorem.

For the error $z = (z_1, \ldots, z_n)$, $z_k = u(t_k) - y_k$ we have the relations

$$z_k = S_{k,k-1}z_{k-1} + \sum_{p=0}^{n} \alpha_{kp}z_p + \psi_k, \quad k = 1, \ldots, n, \tag{3.199}$$

where

$$\psi_k = \int_{t_{k-1}}^{t_k} e^{-A_k(t_k-\eta)} \int_{t_{k-1}}^{\eta} e^{-A_k(\eta-s)}[A_k - A(s)][u(s) - P_{n-1}(s;u)]dsd\eta. \tag{3.200}$$

We introduce the matrix

$$\tilde{S} = \{\tilde{s}_{i,k}\}_{i,k=1}^{n} = \begin{pmatrix} I & 0 & 0 & \cdot & \cdot & & 0 & 0 \\ -\tilde{S}_{21} & I & 0 & \cdot & \cdot & & 0 & 0 \\ 0 & -\tilde{S}_{32} & I & \cdot & \cdot & & 0 & 0 \\ & & & \cdot & & & & \\ \cdot & & & & \cdot & & & \cdot \\ 0 & 0 & 0 & \cdot & \cdot & & -\tilde{S}_{n,n-1} & I \end{pmatrix}, \tag{3.201}$$

with $\tilde{S}_{k,k-1} = A_k^\gamma S_{k,k-1} A_{k-1}^{-\gamma}$, the matrix $C = \{\tilde{\alpha}_{k,p}\}_{k,p=1}^n$ with $\tilde{\alpha}_{k,p} = A_k^\gamma \alpha_{k,p} A_p^{-\gamma}$ and the vectors

$$
y = \begin{pmatrix} A_1^\gamma y_1 \\ \cdot \\ \cdot \\ \cdot \\ A_n^\gamma y_n \end{pmatrix}, \quad f = \begin{pmatrix} A_1^\gamma \phi_1 \\ \cdot \\ \cdot \\ \cdot \\ a_n^\gamma \phi_n \end{pmatrix}, \quad \psi = \begin{pmatrix} A_1^\gamma \psi_1 \\ \cdot \\ \cdot \\ \cdot \\ A_n^\gamma \psi_n \end{pmatrix}, \quad \tilde{f} = \begin{pmatrix} A_1^\gamma S_{21} u_0 \\ 0 \\ \cdot \\ \cdot \\ \cdot \\ 0 \end{pmatrix}. \tag{3.202}
$$

It is easy to check that for

$$
\tilde{S}^{-1} = \{\tilde{s}_{i,k}^{-1}\}_{i,k=1}^n
$$

$$
= \begin{pmatrix} I & 0 & 0 & \cdots & 0 & 0 \\ \tilde{S}_{21} & I & 0 & \cdots & 0 & 0 \\ \tilde{S}_{32}\tilde{S}_{21} & \tilde{S}_{32} & I & \cdots & 0 & 0 \\ \cdot & \cdot & \cdot & \cdots & \cdot & \cdot \\ \tilde{S}_{n,n-1}\cdots\tilde{S}_{21} & \tilde{S}_{n,n-1}\cdots\tilde{S}_{32} & \tilde{S}_{n,n-1}\cdots\tilde{S}_{43} & \cdots & \tilde{S}_{n,n-1} & I \end{pmatrix} \tag{3.203}
$$

we have

$$
\tilde{S}^{-1}\tilde{S} = \begin{pmatrix} I & 0 & \cdots & 0 \\ 0 & I & \cdots & 0 \\ \cdot & \cdot & \cdots & \cdot \\ 0 & 0 & \cdots & I \end{pmatrix}. \tag{3.204}
$$

Remark 3.23. Using results of [16], one can get a parallel and sparse approximation of operator exponentials in $\tilde{S}^{-1}$ and, as a consequence, a parallel and sparse approximation of $\tilde{S}^{-1}$.

We get from (3.192), (3.199)

$$
A_k^\gamma y_k = \tilde{S}_{k,k-1} A_{k-1}^\gamma y_{k-1} + \sum_{p=0}^n \tilde{\alpha}_{kp} A_p^\gamma y_p + A_k^\gamma \phi_k,
$$

$$
A_k^\gamma z_k = \tilde{S}_{k,k-1} A_{k-1}^\gamma z_{k-1} + \sum_{p=0}^n \tilde{\alpha}_{kp} A_p^\gamma z_p + A_k^\gamma \psi_k, \tag{3.205}
$$

or in matrix form

$$
\tilde{S}y = Cy + f + \tilde{f},
$$

$$
\tilde{S}z = Cz + \psi \tag{3.206}
$$

with

$$
z = \begin{pmatrix} A_1^\gamma z_1 \\ \cdot \\ \cdot \\ \cdot \\ A_n^\gamma z_n \end{pmatrix}. \tag{3.207}
$$

In the next lemma we estimate the norms of C and $\tilde{S}^{-1}C$.

Lemma 3.24. *The estimates*

$$|\|C\|| \leq c(\gamma, T)n^{2\gamma-4}\ln n, \tag{3.208}$$

$$|\|\tilde{S}^{-1}C\|| \leq c(\gamma, T)n^{2\gamma-3}\ln n \tag{3.209}$$

with a positive constant $c = c(T, \gamma)$ depending on γ and the interval length T but independent of n and such that $c = c(T, \gamma) \to \infty$ as $\gamma \to 1$ hold true.

Proof. Assumption (3.119) together with (3.117),(3.153), (3.150), (3.178) imply

$$
\begin{aligned}
\|\tilde{\alpha}_{kp}\| &= \|A_k^\gamma \alpha_{kp} A_p^{-\gamma}\| \\
&= \|\int_{t_{k-1}}^{t_k} A_k^\gamma e^{-A_k(t_k-\eta)}[A_k - A(\eta)] \\
&\quad \times A_k^{-\gamma}\int_{t_{k-1}}^\eta A_k^\gamma e^{-A_k(\eta-s)}[A_k - A(s)]A_p^{-\gamma}L_{p,n-1}(\eta)d\eta\| \\
&\leq \left(1 + \tilde{L}_\gamma T\right)\left(c_\gamma \tilde{L}_{1,\gamma}\right)^2 \int_{t_{k-1}}^{t_k} |t_k - \eta|^{1-\gamma} \\
&\quad \times \int_{t_{k-1}}^\eta |\eta - s|^{-\gamma}|t_k - s||L_{p,n-1}(s)|dsd\eta.
\end{aligned}
\tag{3.210}
$$

Due to (3.210) we have

$$
\begin{aligned}
|\|C\|| &= \max_{1 \leq k \leq n} \sum_{p=1}^n \|\tilde{\alpha}_{kp}\| \\
&\leq \left(1 + \tilde{L}_\gamma T\right)\left(c_\gamma \tilde{L}_{1,\gamma}\right)^2 \Lambda_n \max_{1 \leq k \leq n} \int_{t_{k-1}}^{t_k} |t_k - \eta|^{1-\gamma} \\
&\quad \times \int_{t_{k-1}}^\eta |\eta - s|^{-\gamma}|t_k - s|dsd\eta \\
&\leq \left(1 + \tilde{L}_\gamma T\right)\left(c_\gamma \tilde{L}_{1,\gamma}\right)^2 \Lambda_n \max_{1 \leq k \leq n} \frac{\tau_k}{1-\gamma}\int_{t_{k-1}}^{t_k} |t_k - \eta|^{1-\gamma}|\eta - t_{k-1}|^{1-\gamma}d\eta \\
&\leq \left(1 + \tilde{L}_\gamma T\right)\left(c_\gamma \tilde{L}_{1,\gamma}\right)^2 \Lambda_n \max_{1 \leq k \leq n} \frac{\tau_k^{2-\gamma}}{1-\gamma}\int_{t_{k-1}}^{t_k} |t_k - \eta|^{1-\gamma}d\eta \\
&\leq c(\gamma, T)\Lambda_n \tau_{\max}^{4-2\gamma} \leq c(\gamma, T)n^{2\gamma-4}\ln n \tag{3.211}
\end{aligned}
$$

where $c(\gamma, T) = c\frac{\left(1+\tilde{L}_\gamma T\right)\left(c_\gamma \tilde{L}_{1,\gamma}\right)^2}{(1-\gamma)(2-\gamma)}$, c is a constant independent of n, γ and (3.208) is proved.

Furthermore, the inequalities (3.117), (3.119), (3.120) imply

$$\|\tilde{S}_{k,k-1}\| \le e^{-\omega \tau_k} + c_\gamma \tilde{L}_{1,\gamma}(1 + \tilde{L}_\gamma \tau_k) \int_{t_{k-1}}^{t_k} |t_k - \eta|^{-\gamma} |t_k - \eta| e^{-\omega(\eta - t_{k-1})} d\eta$$

$$\le e^{-\omega \tau_k} \left[1 + \frac{c_\gamma \tilde{L}_{1,\gamma}(1 + \tilde{L}_\gamma \tau_k)}{2 - \gamma} \tau_k^{2-\gamma} \right] \qquad (3.212)$$

which yields

$$\|\tilde{S}^{-1}\| \le \sum_{p=0}^{n-1} q^p = \frac{q^n - 1}{q - 1} \qquad (3.213)$$

with

$$q = \left\{ e^{-\omega \tau_{\max}} \left[1 + \frac{c_\gamma \tilde{L}_{1,\gamma}(1 + \tilde{L}_\gamma \tau_{\max})}{2 - \gamma} \tau_{\max}^{2-\gamma} \right] \right\} \to 1$$

as $\tau_{\max} \to 0$. This means that there exists a constant $C = C(\gamma, c_\gamma, \tilde{L}_\gamma, \tilde{L}_{1,\gamma})$ such that

$$\|\tilde{S}^{-1}\| \le Cn \qquad (3.214)$$

(it is easy to see that $C \le 1$ provided that $-\omega + \frac{c_\gamma \tilde{L}_{1,\gamma}(1 + \tilde{L}_\gamma \tau_{\max})}{2-\gamma} \tau_{\max}^{1-\gamma} \le 0$). This estimate together with (3.211) implies (3.209). The proof is complete. □

Now, we can go to the first main result of this section.

Theorem 3.25. *Let assumptions (3.116)–(3.120) with $\gamma < 1$ hold. Then there exists a positive constant c such that the following hold:*

1. *For n large enough it holds that*

$$\|\|z\|\| \equiv \|\|y - u\|\| \le cn^{2\gamma-3} \ln n E_n(A_0^\gamma u), \quad \gamma \in [0, 1), \qquad (3.215)$$

   where u is the solution of (3.115) and $E_n(A_0^\gamma u)$ is the best approximation of $A_0^\gamma u$ by polynomials of degree not greater than $n - 1$.

2. *The system of linear algebraic equations*

$$Sy = Cy + f \qquad (3.216)$$

   from (3.206) with respect to the approximate solution y can be solved by the fixed-point iteration

$$y^{(k+1)} = S^{-1}Cy^{(k)} + S^{-1}(f - \tilde{f}), \quad k = 0, 1, \ldots; \quad y^{(0)} \text{ arbitrary} \qquad (3.217)$$

   converging at least as a geometrical progression with the denominator $q = c(\gamma, T)n^{2\gamma-3} \ln n < 1$, $\gamma \in [0, 1)$ for n large enough.

Proof. From the second equation in (3.206) we get

$$z = S^{-1}Cz + S^{-1}\psi \tag{3.218}$$

from which due to Lemma 3.19 and (3.213) we get

$$|||z||| \leq cn|||\psi|||. \tag{3.219}$$

Let Π_{n-1} be the set of all polynomials in t with vector coefficients of degree less than or equal to $n-1$. Using the Lebesgue inequality (3.181) the last norm can be estimated as

$$
\begin{aligned}
|||\psi||| &= \max_{1\leq k\leq n} \left\| \int_{t_{k-1}}^{t_k} A_k^\gamma e^{-A_k(t_k-\eta)} [A_k - A(\eta)] \right. \\
&\quad \times \left. \int_{t_{k-1}}^{\eta} e^{-A_k(\eta-s)} [A_k - A(s)] A_0^{-\gamma}(A_0^\gamma u(s) - P_{n-1}(s; A_0^\gamma u))dsd\eta \right\| \\
&\leq (1+\tilde{L}_\gamma T)(c_\gamma L_{1,\gamma})^2 \max_{1\leq k\leq n} \int_{t_{k-1}}^{t_k} |t_k - \eta|^{-\gamma}|t_k - \eta| \\
&\quad \times \int_{t_{k-1}}^{\eta} |\eta - s|^{-\gamma}|t_k - s|\|A_0^\gamma u(\eta) - P_{n-1}(s; A_0^\gamma u)\|dsd\eta \\
&\leq (1+\tilde{L}_\gamma T)(c_\gamma L_{1,\gamma})^2(1 + \Lambda_n)E_n(A_0^\gamma u) \\
&\quad \times \max_{1\leq k\leq n} \left\{ \int_{t_{k-1}}^{t_k} |t_k - \eta|^{-\gamma}|t_k - \eta| \int_{t_{k-1}}^{\eta} |\eta - s|^{-\gamma}|t_k - s|dsd\eta \right\} \\
&\leq cc(\gamma, T)E_n(A_0^\gamma u)n^{2\gamma-4} \ln n
\end{aligned}
\tag{3.220}
$$

and taking into account (3.219), we get the first assertion of the theorem.

The second assertion is a simple consequence of (3.216) and (3.209), which completes the proof of the theorem. □

Under somewhat stronger assumptions on the operator $A(t)$ one can improve the error estimate for our method in the case $0 \leq \gamma \leq 1$. In order to do it we need the following lemma.

Lemma 3.26. *Let $L_{\nu,n-1}(t)$ be the Lagrange fundamental polynomials related to the Chebyshev interpolation nodes (zeros of the Chebyshev polynomial of the first kind $T_n(t)$). Then*

$$\sum_{\nu=1}^{n} |L'_{\nu,n-1}(t)| \leq \frac{1}{\sqrt{1-x^2}} \sqrt{2/3} n^{3/2}. \tag{3.221}$$

Proof. Let $x \in [-1,1]$ be an arbitrary point and let $\epsilon_\nu = \text{sign}(L'_{\nu,n-1}(x))$. We consider the polynomial of t,

$$\rho(t;x) = \sum_{\nu=1}^{n} \epsilon_\nu(x)L_{\nu,n-1}(t) = \sum_{\nu=1}^{n-1} c_\nu(x)T_\nu(t). \tag{3.222}$$

Since $\rho^2(t;x)$ is the polynomial of degree $2n-2$, then using the Gauß-Chebyshev quadrature rule and the property $L_{k,n-1}(t_\nu) = \delta_{k,\nu}$ of the fundamental Lagrange polynomials ($\delta_{k,\nu}$ is the Kronecker symbol), we get

$$\int_{-1}^{1} \frac{\rho^2(t;x)}{\sqrt{1-t^2}} dt = \sum_{\nu=0}^{n-1} c_\nu^2(x) \frac{\pi}{2} = \sum_{\nu=1}^{n} \lambda_\nu \rho^2(t_\nu;x)$$
$$= \sum_{\nu=1}^{n} \lambda_\nu \epsilon_\nu^2 = \sum_{\nu=1}^{n} \lambda_\nu = \int_{-1}^{1} \frac{1}{\sqrt{1-t^2}} dt = \pi \tag{3.223}$$

with the quadrature coefficients λ_ν which yields

$$\sum_{\nu=0}^{n-1} c_\nu^2(x) = 2. \tag{3.224}$$

The next estimate

$$\rho'(x) = \sum_{\nu=1}^{n} |L'_{\nu,n-1}(x)| \le \sum_{\nu=0}^{n-1} |c_\nu||T'_\nu(x)|$$
$$= \sum_{\nu=0}^{n-1} |c_\nu| \frac{\nu}{\sqrt{1-x^2}} \le \frac{1}{\sqrt{1-x^2}} \left(\sum_{\nu=0}^{n-1} c_\nu^2\right)^{1/2} \left(\sum_{\nu=1}^{n-1} \nu^2\right)^{1/2} \tag{3.225}$$
$$\le \frac{1}{\sqrt{1-x^2}} \left(\sum_{\nu=0}^{n-1} (c_\nu)^2\right)^{1/2} \sqrt{n^3/3}$$

together with (3.224) proves the lemma. $\qquad\square$

Now we are in the position to prove the following important result of this section.

Lemma 3.27. *Let the operator $A(t)$ be strongly continuous differentiable on $[0,T]$ (see [38], Ch. 2, 1, p. 218, [37]), satisfy condition (3.119) and $A'(s)A^{-\gamma}(0)$ be bounded for all $s \in [0,T]$ and $\gamma \in [0,1]$ by a constant c'. Then for n large enough the following estimates hold true:*

$$|||C||| \le cn^{\gamma-5/2}, \quad \gamma \in [0,1], \tag{3.226}$$
$$|||\tilde{S}^{-1}C||| \le cn^{\gamma-3/2}, \quad \gamma \in [0,1] \tag{3.227}$$

with some positive constant c independent of n, γ.

Proof. Opposite to the proof of Lemma 3.19 (see (3.210)), we estimate $\tilde{\alpha}_{kp}$ as

$$\|\tilde{\alpha}_{kp}\| = \|A_k^\gamma \alpha_{kp} A_p^{-\gamma}\|$$
$$= \left\| \int_{t_{k-1}}^{t_k} A_k^\gamma e^{-A_k(t_k-\eta)} [A_k - A(\eta)] \right.$$

$$\times A_k^{-1} \int_{t_{k-1}}^{\eta} \frac{de^{-A_k(\eta-s)}}{ds} [A_k - A(s)]A_p^{-\gamma} L_{p,n-1}(\eta) d\eta \|$$

$$= \| \int_{t_{k-1}}^{t_k} A_k^{\gamma} e^{-A_k(t_k-\eta)} [A_k - A(\eta)]A_k^{-1} \big\{ [A_k - A(\eta)]A_p^{-\gamma} L_{p,n-1}(\eta)$$

$$- e^{-A_k(\eta-t_{k-1})} [A_k - A_{k-1}]A_p^{-\gamma} L_{p,n-1}(t_{k-1})$$

$$+ \int_{t_{k-1}}^{\eta} e^{-A_k(\eta-s)} A'(s)A_p^{-\gamma} L_{p,n-1}(s) ds$$

$$- \int_{t_{k-1}}^{\eta} e^{-A_k(\eta-s)} [A_k - A(s)]A_p^{-\gamma} L'_{p,n-1}(s) ds \big\} d\eta \|$$

$$\leq \int_{t_{k-1}}^{t_k} c_{\gamma} \tilde{L}_{1,1} (t_k - \eta)^{-\gamma} (t_k - \eta) \big\{ \tilde{L}_{1,\gamma}(t_k - \eta)(1 + \tilde{L}_{\gamma} T)|L_{p,n-1}(\eta)|$$

$$+ \tilde{L}_{1,\gamma} \tau_k (1 + \tilde{L}_{\gamma} T)\delta_{p,k-1} + c'(1 + \tilde{L}_{\gamma} T) \int_{t_{k-1}}^{\eta} |L_{p,n-1}(s)| ds$$

$$+ \int_{t_{k-1}}^{\eta} \tilde{L}_{1,\gamma}(t_k - s)(1 + \tilde{L}_{\gamma} T)|L'_{p,n-1}(s)| ds \big\} d\eta$$

$$\leq \int_{t_{k-1}}^{t_k} c_{\gamma} \tilde{L}_{1,1} \big\{ \tilde{L}_{1,\gamma}(t_k - \eta)^{2-\gamma}|L_{p,n-1}(\eta)|$$

$$+ \tilde{L}_{1,\gamma} \tau_k (1 + \tilde{L}_{\gamma} T)(t_k - \eta)^{1-\gamma} \delta_{p,k-1}$$

$$+ c'(1 + \tilde{L}_{\gamma} T)(t_k - \eta)^{1-\gamma} \int_{t_{k-1}}^{\eta} |L_{p,n-1}(s)| ds$$

$$+ \tilde{L}_{1,\gamma} \tau_k (1 + \tilde{L}_{\gamma} T)(t_k - \eta)^{1-\gamma} \int_{t_{k-1}}^{\eta} |L'_{p,n-1}(s)| ds \big\} d\eta. \qquad (3.228)$$

Using this inequality together with (3.221), (3.213) and the relations $\arcsin t = \pi/2 - \arccos t$, $t_k = \cos \frac{2k-1}{2n} \pi$, $k = 1, \ldots, n$, we arrive at the estimates

$$\|\|C\|\| = \max_{1 \leq k \leq n} \sum_{p=1}^{n} \|A_k^{\gamma} \alpha_{k,p} A_p^{-\gamma}\|$$

$$\leq M \left\{ n^{\gamma-3} \ln n + n^{\gamma-3} + n^{\gamma-3/2} \max_{1 \leq k \leq n} \int_{t_{k-1}}^{t_k} \frac{1}{\sqrt{1 - s^2}} ds \right\} \qquad (3.229)$$

$$\leq M \left\{ n^{\gamma-3} \ln n + n^{\gamma-3} + n^{\gamma-3/2} \max_{1 \leq k \leq n} (\arcsin t_k - \arcsin t_{k-1}) \right\}$$

$$\leq M n^{\gamma-5/2} \|\|\tilde{S}^{-1} C\|\| \leq M n^{\gamma-3/2}$$

with a constant M independent of n. The proof is complete. $\square$

Remark 3.28. If an operator $A(t)$ is strongly continuous differentiable, then condition (3.119) holds true with $\gamma = 1$ and the operator $A(t)A^{-1}(0)$ is uniformly bounded (see [38], Ch. 2, 1, p. 219).

Now, we can go to the second main result of this subsection.

Theorem 3.29. *Let the assumptions of Lemma 3.27 and conditions (3.116)–(3.120) hold. Then there exists a positive constant c such that the following statements hold.*

1. *For $\gamma \in [0,1)$ and n large enough it holds that*

$$|||z||| \equiv |||y - u||| \leq cn^{2\gamma-3} \ln n E_n(A_0^\gamma u), \qquad (3.230)$$

   where u is the solution of (3.115) and $E_n(A_0^\gamma u)$ is the best approximation of $A_0^\gamma u$ by polynomials of degree not greater than $n - 1$.

2. *The system of linear algebraic equations (3.216) with respect to the approximate solution y can be solved by the fixed-point iteration*

$$y^{(k+1)} = S^{-1}Cy^{(k)} + S^{-1}f, \quad k = 0, 1, \ldots; \quad y^{(0)} \text{ arbitrary} \quad (3.231)$$

   converging at least as a geometrical progression with the denominator $q = c(\gamma, T)n^{\gamma-3/2} < 1$ for n large enough.

Proof. Proceeding analogously as in the proof of Theorem 3.25 and using Lemma 3.27 and (3.213), we get

$$|||z||| \leq cn|||\psi|||. \qquad (3.232)$$

For the norm $|||\psi|||$, we have (see (3.220))

$$|||\psi||| \leq c(1 + \tilde{L}_\gamma T)E_n(A_0^\gamma u)n^{2\gamma-4} \ln n, \quad \gamma \in [0,1) \qquad (3.233)$$

which together with (3.232) leads to the estimate (3.230) and to the first assertion of the theorem.

The second assertion is a consequence of (3.229). The proof is complete. □

Remark 3.30. A simple generalization of Bernstein's theorem (see [50, 51, 52]) to vector-valued functions gives the estimate

$$E_n(A_0^\gamma u) \leq \rho_0^{-n} \qquad (3.234)$$

for the value of the best polynomial approximation provided that $A_0^\gamma u$ can be analytically extended from $[-1,1]$ into an ellipse with the focus at the points $+1, -1$ and with the sum of semi-axes $\rho_0 > 1$.

If $A_0^\gamma u$ is p times continuously differentiable, then a generalization of Jackson's theorem (see [50, 51, 52]) gives

$$E_n(A_0^\gamma u) \leq c_p n^{-p} \omega\left(\frac{d^p A_0^\gamma u}{dt^p}; n^{-1}\right) \qquad (3.235)$$

with the continuity modulus ω.

Further generalizations for the Sobolev spaces of the vector-valued functions can be proven analogously [7], Ch. 9 (see also section 2.1). Let us define the weighted Banach space of vector-valued functions $L_w^p(-1,1)$, $1 \le p \le +\infty$ with the norm

$$\|u\|_{L_w^p(-1,1)} = \left(\int_{-1}^{1} \|u(t)\|^p w(t) dt \right)^{1/p} \tag{3.236}$$

for $1 \le p < \infty$ and

$$\|u\|_{L^\infty(-1,1)} = \sup_{t \in (-1,1)} \|u(t)\| \tag{3.237}$$

for $p = \infty$. The weighted Sobolev space is defined by

$$H_w^m(-1,1) = \left\{ v \in L_w^2(-1,1) : \text{ for } 0 \le k \le m, \; \frac{d^k v}{dt^k} \in L_w^2(-1,1) \right\}$$

with the norm

$$\|u\|_{H_w^m(-1,1)} = \left(\sum_{k=0}^{m} \left\| \frac{d^k v}{dt^k} \right\|_{L_w^2(-1,1)}^2 \right)^{1/2}. \tag{3.238}$$

Then one gets for the Chebyshev weight $w(t) = \frac{1}{\sqrt{1-t^2}}$ (see [7], p. 295–298), for the polynomial of the best approximation $B_n(t)$ and for the interpolation polynomial $P_n(t)$ with the Gauss (roots of the Chebyshev polynomial $T_{n+1}(t)$), Gauss-Radau (roots of the polynomial $T_{n+1}(t) - \frac{T_{n+1}(-1)}{T_n(1)} T_n(t)$) or the Gauss-Lobatto (roots of the polynomial $p(t) = T_{n+1}(t) + aT_n(t) + bT_{n-1}(t)$ with a, b such that $p(-1) = p(1) = 0$) nodes

$$E_n(u) \equiv \|u - B_n u\|_{L^\infty(-1,1)} \le cn^{1/2-m} \|u\|_{H_w^m(-1,1)},$$
$$\|u - B_n u\|_{L_w^2(-1,1)} \le cn^{-m} \|u\|_{H_w^m(-1,1)},$$
$$\|u - P_n u\|_{L_w^2(-1,1)} \le cn^{-m} \|u\|_{H_w^m(-1,1)}, \tag{3.239}$$
$$\|u' - (P_n u)'\|_{L_w^2(-1,1)} \le cn^{2-m} \|u\|_{H_w^m(-1,1)}.$$

When the function u is analytic in $[-1,1]$ and has a regularity ellipse whose sum of semi-axes equals e^{η_0}, then

$$\|u' - (P_n u)'\|_{L_w^2(-1,1)} \le c(\eta) n^2 e^{-n\eta} \quad \forall \eta \in (0, \eta_0). \tag{3.240}$$

For the Legendre weight $w(t) = 1$ one has (see [7], p. 289–294)

$$\|u - B_n u\|_{L^p(-1,1)} \le cn^{-m} \|u\|_{H^m(-1,1)}, \quad 2 < p \le \infty,$$
$$\|u - B_n u\|_{H^l(-1,1)} \le cn^{2l-m+1/2} \|u\|_{H^m(-1,1)}, \quad 1 \le l \le m,$$
$$\|u - P_n u\|_{H^l(-1,1)} \le cn^{2l-m+1/2} \|u\|_{H^m(-1,1)}, \quad 1 \le l \le m, \tag{3.241}$$
$$\|u' - (P_n u)'\|_{L^2(-1,1)} \le cn^{5/2-m} \|u\|_{H_w^m(-1,1)},$$

where the interpolation polynomial $P_n(t)$ can be taken with the Gauss (roots of the Legendre polynomial $L_{n+1}(t)$), Gauss-Radau (roots of the polynomial $L_{n+1}(t) - \frac{L_{n+1}(-1)}{L_n(1)} L_n(t)$) or the Gauss-Lobatto (roots of the polynomial $p(t) = L_{n+1}(t) + aL_n(t) + bL_{n-1}(t)$ with a, b such that $p(-1) = p(1) = 0$) nodes.

Note that the restriction $\gamma \neq 1$ in Theorem 3.29 is only due to the estimate (3.220). Below we show how this restriction can be removed.

Using (3.178), (3.239), we estimate the norm $\||\psi|\|_2$ for $\gamma = 1$ as

$$\||\psi|\| = \max_{1 \leq k \leq n} \| \int_{t_{k-1}}^{t_k} A_k^\gamma e^{-A_k(t_k - \eta)} (A_k - A(\eta))$$

$$\times \int_{t_{k-1}}^{\eta} e^{-A_k(\eta - s)} (A_k - A(s)) A_k^{-\gamma}(A_k^\gamma A_0^{-\gamma})(A_0^\gamma u(s) - P_{n-1}(s; A_0^\gamma u)) ds d\eta\|$$

$$= \max_{1 \leq k \leq n} \| \int_{t_{k-1}}^{t_k} A_k e^{-A_k(t_k - \eta)} (A_k - A(\eta)) A_k^{-1}$$

$$\times \int_{t_{k-1}}^{\eta} \frac{d e^{-A_k(\eta - s)}}{ds} (A_k - A(s)) A_k^{-1}(A_k A_0^{-1})(A_0 u(s) - P_{n-1}(s; A_0 u)) ds d\eta\|$$

$$= \max_{1 \leq k \leq n} \| \int_{t_{k-1}}^{t_k} A_k e^{-A_k(t_k - \eta)} (A_k - A(\eta)) A_k^{-1}$$

$$\times \left[(A_k - A(\eta)) A_k^{-1}(A_k A_0^{-1})(A_0 u(\eta) - P_{n-1}(\eta; A_0 u)) \right.$$

$$+ \int_{t_{k-1}}^{\eta} e^{-A_k(\eta - s)} A'(s) A_k^{-1}(A_k A_0^{-1})(A_0 u(s) - P_{n-1}(s; A_0 u)) ds$$

$$\left. - \int_{t_{k-1}}^{\eta} e^{-A_k(\eta - s)} (A_k - A(s)) A_k^{-1}(A_k A_0^{-1})(A_0 u'(s) - (P_{n-1}(s; A_0 u))') ds \right] d\eta\|$$

$$\leq c_1 \tilde{L}_{1,1} \max_{1 \leq k \leq n} \int_{t_{k-1}}^{t_k} |t_k - \eta|^{-1}|t_k - \eta|$$

$$\times \left[|t_k - \eta|(1 + \tilde{L}_1 T) \max_{\eta \in [-1,1]} \|(A_0 u(\eta) - P_{n-1}(\eta; A_0 u))\| \right.$$

$$+ c'(1 + \tilde{L}_1 T) \int_{t_{k-1}}^{\eta} \|(A_0 u(s) - P_{n-1}(s; A_0 u))\| ds$$

$$\left. + (1 + \tilde{L}_1 T) \int_{t_{k-1}}^{\eta} |t_k - s| \|(A_0 u'(s) - (P_{n-1}(s; A_0 u))')\| ds \right] d\eta$$

$$\leq c_1 \tilde{L}_{1,1} \max_{1 \leq k \leq n} \left[\tau_k^2(1 + \tilde{L}_1 T)(1 + \Lambda_n) E_n(A_0 u) + c'(1 + \tilde{L}_1 T)\tau_k^2(1 + \Lambda_n) E_n(A_0 u) \right.$$

$$\left. + (1 + \tilde{L}_1 T)\tau_k^2 \int_{t_{k-1}}^{t_k} \|(A_0 u'(s) - (P_{n-1}(s; A_0 u))')\| ds \right]$$

$$\leq c_1 \tilde{L}_{1,1} \max_{1 \leq k \leq n} \left[\tau_k^2 (1 + \tilde{L}_1 T)(1 + \Lambda_n) E_n(A_0 u) + c'(1 + \tilde{L}_1 T) \tau_k^2 (1 + \Lambda_n) E_n(A_0 u) \right.$$

$$+ (1 + \tilde{L}_1 T) \tau_k^2 \left(\int_{t_{k-1}}^{t_k} \frac{1}{\sqrt{1 - s^2}} ds \right)^{1/2}$$

$$\left. \times \left(\int_{t_{k-1}}^{t_k} \frac{1}{\sqrt{1 - s^2}} \|(A_0 u'(s) - (P_{n-1}(s; A_0 u))')\| ds \right)^{1/2} \right]$$

$$\leq c \left[n^{-2} \ln n E_n(A_0 u) + n^{-5/2} \|u' - (P_{n-1} u)'\|_{L^2_w(-1,1)} \right]$$

$$\leq c(n^{-(m+2)} \ln n + n^{-(m+1/2)}) \|u\|_{H^m_w(-1,1)} \leq c n^{-(m+1/2)} \|u\|_{H^m_w(-1,1)}$$

provided that the solution u of problem (3.115) belongs to the Sobolev class $H^m_w(-1,1)$. If u is analytic in $[-1,1]$ and has a regularity ellipse with the sum of the semi-axes equal to $e^{\eta_0} > 1$, then using (3.240), we get

$$\||\psi\|| \leq c(\eta_0) n^2 e^{-n\eta_0}. \tag{3.242}$$

Now, Lemma 3.27 together with the last estimates for $\||\psi\||$ yields the following third main result of this section.

Theorem 3.31. *Let the assumptions of Lemma 3.27 and conditions* (3.116)–(3.120) *with* $\gamma = 1$ *hold. Then there exists a positive constant* c *such that the following statements hold.*

1. *For* $\gamma = 1$ *and* n *large enough we have*

$$\||z\|| \equiv \||y - u\|| \leq c n^{-m} \|u\|_{H^m_w(-1,1)} \tag{3.243}$$

   provided that the solution u *of problem* (3.115) *belongs to the class* $H^m_w(-1,1)$ *with* $w(t) = \frac{1}{\sqrt{1-t^2}}$.

2. *For* $\gamma = 1$ *and* n *large enough it holds that*

$$\||z\|| \equiv \||y - u\|| \leq c(\eta_0) n^{3/2} e^{-n\eta_0} \tag{3.244}$$

   provided that u *is analytic in* $[-1,1]$ *and has a regularity ellipse with the sum of the semi-axes equal to* $e^{\eta_0} > 1$.

3. *The system of linear algebraic equations* (3.216) *with respect to the approximate solution* y *can be solved by the fixed-point iteration*

$$y^{(k+1)} = S^{-1} C y^{(k)} + S^{-1} f, \quad k = 0, 1, \ldots; \quad y^{(0)} \text{ arbitrary} \tag{3.245}$$

   converging at least as a geometrical progression with the denominator $q = c n^{-1/2} < 1$ *for* n *large enough.*

Remark 3.32. Using estimates (3.241), one can analogously construct a discrete scheme on the Gauss, the Gauss-Radau or the Gauss-Lobatto grids relative to $w(t) = 1$ (i.e., connected with the Legendre orthogonal polynomials) and get the corresponding estimates in the $L^2(-1,1)$-norm.

3.3 An exponentially convergent algorithm for nonlinear differential equations in Banach space

3.3.1 Introduction

We consider the problem

$$\frac{\partial u(t)}{\partial t} + Au(t) = f(t, u(t)), \quad t \in (0, 1],$$

$$u(0) = u_0,$$

(3.246)

where $u(t)$ is an unknown vector-valued function with values in a Banach space X, $u_0 \in X$ is a given vector, $f(t, u) : (\mathbb{R}_+ \times X) \to X$ is a given function (nonlinear operator) and A is a linear densely defined closed operator with the domain $D(A)$ acting in X. The abstract setting (3.246) covers many applied problems such as nonlinear heat conduction or diffusion in porous media, the flow of electrons and holes in semiconductors, nerve axon equations, chemically reacting systems, equations of population genetics theory, dynamics of nuclear reactors, Navier-Stokes equations of viscous flow etc. (see, e.g., [32] and the references therein). This fact together with theoretical interest is an important reason to study efficient discrete approximations of problem (3.246).

In this section we construct exponentially convergent approximations to the solution of nonlinear problem (3.246). To this end we use an equivalent Volterra integral equation including the operator exponential and represent the operator exponential by a Dunford-Cauchy integral along a hyperbola enveloping the spectrum of the operator coefficient. Then we approximate the integrals involved using Chebyshev interpolation and an appropriate Sinc quadrature.

Problem (3.246) is equivalent to the nonlinear Volterra integral equation

$$u(t) = u_h(t) + u_{nl}(t),$$

(3.247)

where

$$u_h(t) = T(t)u_0,$$

(3.248)

$T(t) = e^{-At}$ is the operator exponential (the semi-group) generated by A and the nonlinear term is given by

$$u_{nl}(t) = \int_0^t e^{-A(t-s)} f(s, u(s)) ds.$$

(3.249)

We suppose that the solution $u(t)$ and the function $f(t, u(t))$ can be analytically extended (with respect to t) into a domain which we will describe below.

3.3.2 A discretization scheme of Chebyshev type

Changing in (3.247) the variables by

$$t = \frac{x + 1}{2},$$

(3.250)

we transform problem (3.247) to the following problem on the interval $[-1, 1]$:

$$u(\frac{x+1}{2}) = g_h(x) + g_{nl}(x, u) \tag{3.251}$$

with

$$\begin{aligned} g_h(x) &= e^{-A\frac{x+1}{2}} u_0, \\ g_{nl}(x, u) &= \frac{1}{2} \int_{-1}^{x} e^{-A\frac{x-\xi}{2}} f(\frac{\xi+1}{2}, u(\frac{\xi+1}{2})) d\xi. \end{aligned} \tag{3.252}$$

Using the representation of the operator exponential by the Dunford-Cauchy integral along the integration path Γ_I defined above in (3.33), (3.42) and enveloping the spectral curve Γ_0 we obtain

$$\begin{aligned} g_h(x) &= e^{-A\frac{x+1}{2}} u_0 = \frac{1}{2\pi i} \int_{\Gamma_I} e^{-z\frac{x+1}{2}} [(zI - A)^{-1} - \frac{1}{z}I] u_0 dz, \\ g_{nl}(x, u) &= \frac{1}{2} \int_{-1}^{x} e^{-A\frac{x-\eta}{2}} f(\frac{\eta+1}{2}, u(\frac{\eta+1}{2})) d\eta \\ &= \frac{1}{4\pi i} \int_{-1}^{x} \int_{\Gamma_I} e^{-z\frac{x-\eta}{2}} [(zI - A)^{-1} - \frac{1}{z}I] f(\frac{\eta+1}{2}, u(\frac{\eta+1}{2})) dz d\eta \end{aligned} \tag{3.253}$$

(note, that $P.V. \int_{\Gamma_I} z^{-1} dz = 0$ but this term in the resolvent provides the numerical stability of the algorithm below when $t \to 0$, see section 3.1 or [27] for details). After parametrizing the first integral in (3.253) by (3.42) we have

$$g_h(x) = \frac{1}{2\pi i} \int_{-\infty}^{\infty} \mathcal{F}_h(x, \xi) d\xi \tag{3.254}$$

with

$$\mathcal{F}_h(x, \xi) = F_A((x+1)/2, \xi) u_0 \tag{3.255}$$

(in the case $A = 0$ we define $F_A(t, \xi) = 0$).

We approximate integral (3.254) by the following Sinc-quadrature (see (3.50), (3.53), (3.54))

$$g_{h,N_1}(x) = \frac{h}{2\pi i} \sum_{k=-N_1}^{N_1} \mathcal{F}_h(x, kh), \quad h = \sqrt{\frac{2\pi d}{\alpha(N_1 + 1)}} \tag{3.256}$$

with the error

$$\|\eta_{N_1}(\mathcal{F}_h, h)\| = \|\mathcal{E}((x+1)/2) u_0\| \leq \frac{c}{\alpha} exp\left(-\sqrt{\frac{\pi d\alpha}{2}(N_1 + 1)}\right) \|A^\alpha u_0\|, \tag{3.257}$$

where

$$\mathcal{E}((x-\eta)/2) = \frac{1}{2\pi i} \int_{-\infty}^{\infty} F_A((x-\eta)/2, \xi)d\xi - \frac{1}{2\pi i} \sum_{k=-N_1}^{N_1} F_A((x-\eta)/2, kh) \quad (3.258)$$

and the constant c is independent of x, N_1. Analogously, we transform the second integral in (3.253) to

$$g_{nl}(x, u) = \frac{1}{4\pi i} \int_{-1}^{x} \int_{\Gamma_I} e^{-z\frac{x-\eta}{2}} [(zI - A)^{-1} - \frac{1}{z}I] f(\frac{\eta+1}{2}, u(\frac{\eta+1}{2})) dz d\eta$$

$$= \frac{1}{4\pi i} \int_{-1}^{x} \int_{-\infty}^{\infty} \mathcal{F}_{nl}(x, \xi, \eta) d\xi d\eta,$$

$$(3.259)$$

where

$$\mathcal{F}_{nl}(x, \xi, \eta) = F_A((x-\eta)/2, \xi) f(\frac{\eta+1}{2}, u(\frac{\eta+1}{2})). \quad (3.260)$$

Replacing the infinite integral by quadrature rule (3.256), we arrive at the approximation

$$g_{nl,N_1}(x, u) = \frac{h}{4\pi i} \int_{-1}^{x} \sum_{k=-N_1}^{N_1} \mathcal{F}_{nl}(x, kh, \eta) d\eta. \quad (3.261)$$

To approximate the nonlinear operator $g_{nl,N_1}(x, u)$ we choose the mesh $\omega_N = \{x_{k,N} = \cos\frac{(2k-1)\pi}{2N}, \ k = 1, \ldots, N\}$ on $[-1, 1]$, where $x_{k,N}$ are zeros of the Chebyshev orthogonal polynomial of first kind $T_N(x) = \cos(N \arccos x)$. For the step-sizes $\tau_{k,N} = x_{k,N} - x_{k-1,N}$ it is well known that (see [63], Ch. 6, Th. 6.11.12, [64], p. 123)

$$\tau_{k,N} = x_{k+1,N} - x_{k,N} < \frac{\pi}{N}, k = 1, \ldots, N,$$

$$\tau_{max} = \max_{1 \le k \le N} \tau_{k,N} < \frac{\pi}{N}. \quad (3.262)$$

Let

$$P_{N-1}(x; f(\cdot, u)) = \sum_{p=1}^{N} f((x_{p,N} + 1)/2, u((x_{p,N} + 1)/2)) L_{p,N-1}(x) \quad (3.263)$$

be the interpolation polynomial for the function $f(x, u(x))$ on the mesh ω_N, i.e., $P_{N-1}(x_{k,N}; f(\cdot, u)) = f((x_{k,N} + 1)/2, u((x_{k,N} + 1)/2)), \ k = 1, 2, \ldots, N$, where $L_{p,N-1} = \frac{T_N(x)}{T'_N(x_{p,N})(x-x_{p,N})}, \ p = 1, \ldots, N$ are the Lagrange fundamental polynomials. Given a vector $y = (y_1, \ldots, y_N), \ y_i \in X$ let

$$P_{N-1}(x; f(\cdot, y)) = \sum_{p=1}^{N} f((x_{p,N} + 1)/2, y_p) L_{p,N-1}(x) \quad (3.264)$$

be the polynomial which interpolates $f(x, y)$, i.e., $P_{N-1}(x_{k,N}; f(\cdot, y)) = f((x_{k,N} + 1)/2, y_k))$, $k = 1, 2, \ldots, N$. Substituting $P_{N-1}(t; f(\cdot, y))$ instead of $f(t, u)$ into (3.260), (3.261), we get the approximation

$$g_{nl,N,N_1}(x, y) = \frac{h}{4\pi i} \int_{-1}^{x} \sum_{k=-N_1}^{N_1} F_A((x - \eta)/2, kh) P_{N-1}(\eta; f(\cdot, y)) d\eta. \quad (3.265)$$

Substituting approximations (3.256) and (3.265) into (3.256) and collocating the resulting equation on the grid ω_N we arrive at the following **Algorithm A1** for solving problem (3.251): find $y = (y_1, \ldots, y_N)$, $y_i \in X$ such that

$$y_j = g_{h,N_1}(x_{j,N}) + g_{nl,N,N_1}(x_{j,N}, y), j = 1, \ldots, N \quad (3.266)$$

or

$$\begin{aligned} y_j = \; & \frac{h}{2\pi i} \sum_{k=-N_1}^{N_1} \mathcal{F}_h(x_{j,N}, kh) \\ & + \frac{h}{4\pi i} \sum_{k=-N_1}^{N_1} \int_{-1}^{x_{j,N}} F_A((x_{j,N} - \eta)/2, kh) P_{N-1}(\eta; f(\cdot, y)) d\eta, \\ & j = 1, \ldots, N. \end{aligned} \quad (3.267)$$

Equations (3.266) or (3.267) define a nonlinear operator $\mathcal{A}$ so that

$$y = \mathcal{A}(y) + \phi, \quad (3.268)$$

where

$$\begin{aligned} y = \; & (y_1, y_2, \ldots, y_N), \quad y_i \in X, \\ [\mathcal{A}(y)]_j = \; & \frac{h}{4\pi i} \sum_{k=-N_1}^{N_1} \int_{-1}^{x_{j,N}} F_A((x_{j,N} - \eta)/2, kh) P_{N-1}(\eta; f(\cdot, y)) d\eta, \\ (\phi)_j = \; & \frac{h}{2\pi i} \sum_{k=-N_1}^{N_1} \mathcal{F}_h(x_{j,N}, kh) = \frac{h}{2\pi i} \sum_{k=-N_1}^{N_1} F_A((x_{j,N} + 1)/2, kh) u_0, \\ & j = 1, \ldots, N. \end{aligned} \quad (3.269)$$

This is a system of nonlinear equations which can be solved by an iteration method. Since the integrands in

$$\begin{aligned} I_{j,k} = \; & \int_{-1}^{x_{j,N}} F_A((x_{j,N} - \eta)/2, kh) P_{N-1}(\eta; f(\cdot, y)) d\eta, \\ & j = 1, \ldots, N, \; k = -N_1, \ldots, N_1 \end{aligned}$$

are products of the exponential function and polynomials, these integrals can be calculated analytically, for example, by a computer algebra tools.

For a given vector $y = (y_1, \ldots, y_N)$ the interpolation polynomial $\tilde{u}(x) = P_{N-1}(x; y)$ represents an approximation for $u((x+1)/2) = u(t)$, i.e., $u((x+1)/2) = u(t) \approx P_{N-1}(x; y)$.

3.3.3 Error analysis for a small Lipschitz constant

In this section, we investigate the error of algorithm (3.267). The projection of the exact equation (3.251) onto the grid ω_N provides

$$u(t_j) = \mathrm{e}^{-At_j} u_0 + \frac{1}{2} \int_{-1}^{x_{j,N}} \mathrm{e}^{-A(x_{j,N}-\xi)/2} f\left(\frac{1+\xi}{2}, u\left(\frac{1+\xi}{2}\right)\right) d\xi,$$

$$t_j = \frac{1+x_{j,N}}{2}, \quad x_{j,N} = \cos\frac{(2j-1)\pi}{2N}, \quad j = 1, \ldots, N. \tag{3.270}$$

Using equations (3.267) we represent the error of the algorithm in the form

$$Z_j = u(t_j) - y_j$$
$$= \psi_j^{(0)} + \psi_j^{(1)} + \psi_j^{(2)} + \psi_j^{(3)}, \quad j = 1, \ldots, N, \tag{3.271}$$

where

$$\psi_j^{(0)} = \eta_N(\mathcal{F}_h, h)$$

$$= \left\{ \mathrm{e}^{-At_j} - \frac{h}{2\pi i} \sum_{k=-N_1}^{N_1} z'(kh)\mathrm{e}^{-t_j z(kh)} \left[(z(kh)I - A)^{-1} - \frac{1}{z(kh)}I \right] \right\} u_0$$

$$\psi_j^{(1)} = \frac{1}{2} \int_{-1}^{x_{j,N}} f\left(\frac{1+\eta}{2}, u\left(\frac{1+\eta}{2}\right)\right) \left\{ \mathrm{e}^{-A\frac{(x_{j,N}-\eta)}{2}} \right.$$

$$\left. - \frac{h}{2\pi i} \sum_{k=-N_1}^{N_1} z'(kh)\mathrm{e}^{-z(kh)\frac{(x_{j,N}-\eta)}{2}} \left[(z(kh)I - A)^{-1} - \frac{1}{z(kh)}I \right] \right\} d\eta,$$

$$\psi_j^{(2)} = \frac{h}{2\pi i} \sum_{k=-N_1}^{N_1} z'(kh)\frac{1}{2} \int_{-1}^{x_{j,N}} \mathrm{e}^{-z(kh)(x_{j,N}-\eta)/2} \left[(z(kh)I - A)^{-1} - \frac{1}{z(kh)}I \right]$$

$$\times \left[f\left(\frac{1+\eta}{2}, u\left(\frac{1+\eta}{2}\right)\right) - \sum_{l=1}^{N} f(t_l, u(t_l))L_{l,N-1}(\eta) \right] d\eta,$$

$$\psi_j^{(3)} = \frac{h}{2\pi i} \sum_{k=-N_1}^{N_1} z'(kh)\frac{1}{2} \int_{-1}^{x_{j,N}} \mathrm{e}^{-z(kh)(x_{j,N}-\eta)/2} \left[(z(kh)I - A)^{-1} - \frac{1}{z(kh)}I \right]$$

$$\times \left[\sum_{l=1}^{N} [f(t_l, u(t_l)) - f(t_l, y_l)]L_{l,N-1}(\eta) \right] d\eta,$$

$$z(\xi) = a_I \cosh\xi - ib_I \sinh\xi.$$

Using notation as in (3.255), (3.258) we can write down

$$\psi_j^{(0)} = \mathcal{E}((x_{j,N} + 1)/2)u_0,$$

$$\psi_j^{(1)} = \frac{1}{2} \int_{-1}^{x_{j,N}} \mathcal{E}((x_{j,N} - \eta)/2) f(\frac{1+\eta}{2}, u(\frac{1+\eta}{2}))d\eta,$$

$$\psi_j^{(2)} = \frac{1}{2} \int_{-1}^{x_{j,N}} \frac{h}{2\pi i} \sum_{k=-N_1}^{N_1} F_A((x_{j,N} - \eta)/2, kh)$$

$$\times \left[f(\frac{1+\eta}{2}, u(\frac{1+\eta}{2})) - P_{N-1}(\eta; f(\cdot, u(\cdot))) \right] d\eta,$$

$$\psi_j^{(3)} = \frac{1}{2} \int_{-1}^{x_{j,N}} \frac{h}{2\pi i} \sum_{k=-N_1}^{N_1} F_A((x_{j,N} - \eta)/2, kh)$$

$$\times \left[P_{N-1}(\eta; f(\cdot, u(\cdot)) - f(\cdot, y(\cdot))) \right] d\eta$$

$$= [\mathcal{A}(u)]_j - [\mathcal{A}(y)]_j$$

where $u = (u(t_1), \dots, u(t_N))$, $y = (y_1, \dots, y_N)$ and $\psi_j = \psi_j^{(0)} + \psi_j^{(1)} + \psi_j^{(2)}$ is the truncation error.

For the first summand, we have estimate (3.257):

$$\|\psi_j^{(0)}\| \le \frac{c}{\alpha} \exp\left(-\sqrt{\frac{\pi d\alpha}{2}}(N_1 + 1)\right) \|A^\alpha u_0\| \tag{3.272}$$

and obviously the estimate

$$\|A^\alpha \psi_j^{(0)}\| \le \frac{c}{\beta - \alpha} \exp\left(-\sqrt{\frac{\pi d\alpha}{2}}(N_1 + 1)\right) \|A^\beta u_0\| \quad \forall \beta > \alpha > 0. \tag{3.273}$$

In order to estimate $\psi_j^{(1)}$ we assume that

(i) $f(t, u(t)) \in D(A^\alpha) \ \forall \, t \in [0, 1]$ and $\int_0^1 \|A^\alpha f(t, u(t))\| dt < \infty.$

Using this assumption we obtain analogously to (3.272), (3.273)

$$\|\psi_j^{(1)}\| \le \frac{c}{\alpha} \exp\left(-\sqrt{\frac{\pi d\alpha}{2}}(N_1 + 1)\right) \int_0^1 \|A^\alpha f(t, u(t))\| dt \tag{3.274}$$

and

$$\|A^\alpha \psi_j^{(1)}\| \le \frac{c}{\beta - \alpha} \exp\left(-\sqrt{\frac{\pi d\alpha}{2}}(N_1 + 1)\right)$$

$$\times \int_0^1 \|A^\beta f(t, u(t))\| dt \quad \forall \beta > \alpha > 0. \tag{3.275}$$

In order to estimate $\psi_j^{(2)}$ we assume in addition to the assumption **(i)** that

(ii) the vector-valued function $A^\alpha f(\frac{1+\xi}{2}, u(\frac{1+\xi}{2}))$ of ξ can be analytically extended from the interval $B = [-1, 1]$ into the domain $\mathcal{D}_\rho$ enveloped by the so-called Bernstein's regularity ellipse $\mathcal{E}_\rho = \mathcal{E}_\rho(B)$ (with the foci at $z = \pm 1$ and the sum of semi-axes equal to $\rho > 1$):

$$\mathcal{E}_\rho = \left\{ z \in \mathbb{C} : z = \frac{1}{2}\left(\rho e^{i\varphi} + \frac{1}{\rho}e^{-i\varphi} \right) \right\}$$
$$= \left\{ (x, y) : \frac{x^2}{a^2} + \frac{x^2}{a^2} = 1, \ a = \frac{1}{2}\left(\rho + \frac{1}{\rho} \right), \ b = \frac{1}{2}\left(\rho - \frac{1}{\rho} \right) \right\}.$$

Using (2.32) with $m = 0$, the first inequality (3.36) with $\nu = 0$ and the fact that the Lebesque constant for the Chebyshev interpolation process is bounded by $c \ln N$, we obtain

$$\|\psi_j^{(2)}\| \le c \cdot S_{N_1} \cdot \ln N \cdot E_N\left(A^\alpha f\left(\cdot, u\left(\cdot\right)\right) \right) \tag{3.276}$$

where $S_{N_1} = \sum_{k=-N_1}^{N_1} h|z'(kh)|/|z(kh)|^{1+\alpha}$, c is a constant independent of N, N_1, η and $E_N\left(A^\alpha f\left(\cdot, u\left(\cdot\right)\right) \right)$ is the value of the best approximation of $A^\alpha f(t, u(t))$ by polynomials of degree not greater than $N - 1$ in the maximum norm with respect to t. Using the estimate

$$|z(kh)| = \sqrt{a_I^2 \cosh^2\left(kh\right) + b_I^2 \sinh^2\left(kh\right)}$$
$$\ge a_I \cosh\left(kh\right) \ge a_I e^{|kh|}/2 \tag{3.277}$$

the last sum can be estimated by

$$|S_{N_1}| \le \frac{c}{\sqrt{N_1}} \sum_{k=-N_1}^{N_1} e^{-\alpha|k/\sqrt{N_1}|} \le c \int_{-\sqrt{N_1}}^{\sqrt{N_1}} e^{-\alpha t} dt \le c/\alpha. \tag{3.278}$$

Due to assumption **(ii)** we have for the value of the best polynomial approximation [7, 27]

$$E_N\left(A^\alpha f\left(\cdot, u\left(\cdot\right)\right) \right) \le \frac{\rho^{-N}}{1 - \rho} \sup_{z \in D_\rho} \|A^\alpha f(z, u(z))\|$$

which together with (3.276) and (3.278) yields

$$\|\psi_j^{(2)}\| \le \frac{c}{\alpha} \ln N \rho^{-N} \sup_{z \in D_\rho} \|A^\alpha f(z, u(z))\| \tag{3.279}$$

and

$$\|A^\alpha \psi_j^{(2)}\| \le \frac{c}{\beta - \alpha} \ln N \rho^{-N} \sup_{z \in D_\rho} \|A^\beta f(z, u(z))\| \quad \forall \, \beta > \alpha > 0. \tag{3.280}$$

Before we go over to the estimating of $\psi_j^{(3)}$, let us introduce the functions

$$\Lambda_j^{(1)}(\xi) = \sum_{k=1}^{N} \left| \int_{-1}^{\xi} \chi_j(\eta) L_{k,N-1}(\eta) d\eta \right|, \quad j = 1, \ldots, N, \qquad (3.281)$$

and

$$\Lambda_j^{(2)}(\xi) = \sum_{k=1}^{N} \left| \int_{\xi}^{x_{j,N}} \chi_j(\eta) L_{k,N-1}(\eta) d\eta \right|, \quad j = 1, \ldots, N,$$

with some bounded functions $\chi_j(\eta)$:

$$|\chi_j(\eta)| \le \kappa_j \quad \forall \eta \in [-1,1], \ \ j = 1, \ldots, N \qquad (3.282)$$

and prove the following auxiliary assertion.

Lemma 3.33. *There holds*

$$\begin{aligned}
\Lambda_j^{(1)}(\xi) &\le \kappa_j \sqrt{\pi(\xi + 1)}, \\
\Lambda_j^{(2)}(\xi) &\le \kappa_j \sqrt{\pi(x_{j,N} - \xi)}, \quad \xi \in (-1, x_{j,N}), \ \ j = 1, \ldots, N.
\end{aligned} \qquad (3.283)$$

Proof. Let

$$\epsilon_{k,j}^{(1)} = \mathrm{sign} \left\{ \int_{-1}^{\xi} \chi_j(\eta) L_{k,N-1}(\eta) d\eta \right\},$$

$$\epsilon_{k,j}^{(2)} = \mathrm{sign} \left\{ \int_{\xi}^{x_{j,N}} \chi_j(\eta) L_{k,N-1}(\eta) d\eta \right\},$$

then taking into account that all coefficients of the Gauss quadrature relating to the Chebyshev orthogonal polynomials of first kind are equal to π/N and the Lagrange fundamental polynomials $L_{k,N}(\eta)$ are orthogonal [63] with the weight $1/\sqrt{1-\eta^2}$, we obtain

$$\Lambda_j^{(1)}(\xi) = \int_{-1}^{\xi} \chi_j(\eta) \sum_{k=1}^{N} \epsilon_{k,j}^{(1)} L_{k,N-1}(\eta) d\eta$$

$$\le \sqrt{\xi + 1} \left\{ \int_{-1}^{\xi} \chi_j^2(\eta) \left[\sum_{k=1}^{N} \epsilon_{k,j}^{(1)} L_{k,N-1}(\eta) \right]^2 d\eta \right\}^{1/2}$$

$$\le \kappa_j \sqrt{\xi + 1} \left\{ \int_{-1}^{\xi} \left[\sum_{k=1}^{N} \epsilon_{k,j}^{(1)} L_{k,N-1}(\eta) \right]^2 / \sqrt{1 - \eta^2} d\eta \right\}^{1/2}$$

$$\le \kappa_j \sqrt{\xi + 1} \left\{ \sum_{k,p=1}^{N} \epsilon_{k,j}^{(1)} \epsilon_{p,j}^{(1)} \int_{-1}^{1} L_{k,N-1}(\eta) L_{p,N-1}(\eta) / \sqrt{1 - \eta^2} d\eta \right\}^{1/2}$$

$$= \kappa_j \sqrt{\xi+1} \left\{ \sum_{k=1}^{N} \left(\epsilon_{k,j}^{(1)}\right)^2 \int_{-1}^{1} L_{k,N-1}^2(\eta)/\sqrt{1-\eta^2}d\eta \right\}^{1/2}$$

$$= \kappa_j \sqrt{\pi/N} \sqrt{\xi+1} \left\{ \sum_{k=1}^{N} 1 \right\}^{1/2}$$

$$= \kappa_j \sqrt{\pi(1+\xi)}.$$

Analogously we obtain

$$\Lambda_j^{(2)}(\xi) = \int_{\xi}^{x_{j,N}} \chi_j(\eta) \sum_{k=1}^{N} \epsilon_{k,j}^{(2)} L_{k,N-1}(\eta) d\eta$$

$$\leq \sqrt{x_{j,N}-\xi} \left\{ \int_{\xi}^{x_{j,N}} \chi_j^2(\eta) \left[\sum_{k=1}^{N} \epsilon_{k,j}^{(2)} L_{k,N-1}(\eta) \right]^2 d\eta \right\}^{1/2}$$

$$\leq \kappa_j \sqrt{x_{j,N}-\xi} \left\{ \int_{\xi}^{x_{j,N}} \left[\sum_{k=1}^{N} \epsilon_{k,j}^{(2)} L_{k,N-1}(\eta) \right]^2 /\sqrt{1-\eta^2}d\eta \right\}^{1/2}$$

$$\leq \kappa_j \sqrt{x_{j,N}-\xi} \left\{ \sum_{k,p=1}^{N} \epsilon_{k,j}^{(2)} \epsilon_{p,j}^{(2)} \int_{-1}^{1} L_{k,N-1}(\eta) L_{p,N-1}(\eta)/\sqrt{1-\eta^2}d\eta \right\}^{1/2}$$

$$= \kappa_j \sqrt{\pi(x_{j,N}-\xi)} = \sqrt{2\pi}\kappa_j \cos \frac{(2j-1)\pi}{4N}.$$

The proof is complete. $\qquad\square$

Corollary 3.34. *We define the numbers*

$$\Lambda_j^{(1)} = \Lambda_j^{(1)}(x_{j,N}) = \sum_{k=1}^{N} \left| \int_{-1}^{x_{j,N}} \chi_j(\eta) L_{k,N-1}(\eta) d\eta \right|, \quad j = 1,\dots,N,$$

then using Lemma 3.33 we derive

$$\Lambda_j^{(1)} \leq \kappa_j \sqrt{\pi\left(1+\cos \frac{(2j-1)\pi}{2N}\right)} = \sqrt{2\pi}\kappa_j \cos \frac{(2j-1)\pi}{4N}, \quad j = 1,\dots,N. \quad (3.284)$$

Conjecture. Setting $\chi(\eta) = 1$, we obtain from (3.284) that

$$\Lambda_j^{(1)} \leq \Lambda_{u,j}^{(1)} = \sqrt{2\pi} \cos \frac{(2j-1)\pi}{4N}$$

and the upper bound $\Lambda_{u,j}^{(1)}$ of $\Lambda_j^{(1)}$ is monotonically decreasing in j. In the same time calculations indicate the behavior of $\Lambda_j^{(1)}$ given by Table 3.4.

j	1	2	3	4	5	6	7	8
Λ_j	1.9807	1.8516	1.6037	1.2630	.8934	.5213	.2462	.3097e-1
$\Lambda_{u,j}$	1.9903	1.9138	1.76384	1.5460	1.2687	.9427	.5805	.1960

Table 3.4: The behavior of $\Lambda_j^{(1)}$ and of $\Lambda_{u,j}^{(1)}$ for $N = 8$.

Our hypothesis is that each $\Lambda_j^{(1)}$ is also monotonically decreasing in j and $\Lambda_{u,j}^{(1)} = 2\cos\frac{(2j-1)\pi}{4N}$.

Corollary 3.35. *Numbers $\Lambda_j^{(1)}$ remain bounded also if $\chi_j(\eta) = \mathrm{e}^{-z(x_{j,N}-\eta)}$, where $z = \rho\mathrm{e}^{i\theta}$ is a complex number with $\rho \geq 0$, $\theta \in (-\pi/2, \pi/2)$.*
Actually, we have in this case

$$\Lambda_j^{(1)} = \sum_{k=1}^{N}\left|\int_{-1}^{x_{j,N}} \mathrm{e}^{-\rho\exp\{i\theta\}(x_{j,N}-\eta)}L_{k,N-1}(\eta)d\eta\right|$$

$$\leq \sum_{k=1}^{N}\left|\int_{-1}^{x_{j,N}} \chi_{j,1}(\eta)L_{k,N-1}(\eta)d\eta\right|$$

$$+ \sum_{k=1}^{N}\left|\int_{-1}^{x_{j,N}} \chi_{j,2}(\eta)L_{k,N-1}(\eta)d\eta\right|$$

where

$$\chi_{j,1}(\eta) = \mathrm{e}^{-\rho\cos\theta(x_{j,N}-\eta)}\cos\left[\rho\sin\theta(x_{j,N}-\eta)\right],$$
$$\chi_{j,2}(\eta) = \mathrm{e}^{-\rho\cos\theta(x_{j,N}-\eta)}\sin\left[\rho\sin\theta(x_{j,N}-\eta)\right].$$

Applying Lemma 3.33 for each summand and each of functions $\chi_{j,1}(\eta) \leq 1$ and $\chi_{j,1}(\eta) \leq 1$ we arrive at the estimate

$$\Lambda_j^{(1)} \leq 2\sqrt{2\pi}\cos\frac{(2j-1)\pi}{4N}, \quad j = 1,\dots,N.$$

Now, we are in the position to estimate $\psi_j^{(3)}$. To this end we assume that

(iii) The function $f(t,y) = f(t,y;N)$ in the domain $G = \{(t,y,N) : 0 \leq t \leq 1, |||y-u||| < \gamma, N \geq N_0\}$ in addition to **(i)**, **(ii)** satisfies

$$|||A^{\alpha}[f(t,y_1) - f(t,y_2)]||| \leq L|||y_1 - y_2||| \quad \forall y_1, y_2 \in G,$$

for all $(t,y_i,N) \in G$, $i = 1,2$, where $|||Z||| = |||y-u||| = \max_{j=1,\dots,N}\|y_j - u(t_j)\|$, γ is a positive real constant and N_0 is a fixed natural number large enough.

Under this assumption and taking into account Lemma 3.33, Corollary 3.35 as well as (3.277), (3.278) we have

$$\|\psi_j^{(3)}\| \leq cLh \sum_{k=-N_1}^{N_1} \frac{|z'(kh)|}{|z(kh)|^{1+\alpha}}$$

$$\times \sum_{l=1}^{N} \left| \int_{-1}^{x_{j,N}} e^{-z(kh)(x_{j,N}-\eta)} L_{l,N-1}(\eta)d\eta \right| \|Z_l\| \qquad (3.285)$$

$$\leq \||Z\|| cLh \sum_{k=-N_1}^{N_1} \frac{|z'(kh)|}{|z(kh)|^{1+\alpha}} \Lambda_j^{(1)} \leq cLS_{N_1} \Lambda_j^{(1)} \||Z\|| \leq \frac{c^*}{\alpha} L \||Z\||$$

with a new positive constant c^*. This estimate together with (3.271) implies

$$\||Z\|| \leq \frac{\alpha}{\alpha - c^*L} \||\psi\||$$

with a constant c independent of α, N provided that $c^*L/\alpha < 1$. Analogously, we obtain

$$\|A^\alpha \psi_j^{(3)}\| \leq \frac{c^*}{\beta - \alpha} L \||A^\beta Z\|| \quad \forall \beta > \alpha > 0$$

and

$$\||A^\alpha Z\|| \leq \frac{\beta - \alpha}{\beta - \alpha - c^*L} \||A^\beta \psi\|| \quad \forall \beta > \alpha > 0 \qquad (3.286)$$

with a constant c^* independent of α, β, N provided that $c^*L/(\beta - \alpha) < 1$.

Taking into account (3.271) and estimates (3.272), (3.274), (3.279) as well as (3.285) we arrive at the estimate

$$\||Z\|| \leq \frac{c}{\alpha - c^*L}$$

$$\times \left(e^{-c_1\sqrt{N_1}} \left(\|A^\alpha u_0\| + \int_0^1 \|A^\alpha f(t, u(t))\| dt \right) + \ln N \rho^{-N} \sup_{z \in D_\rho} \|A^\alpha f(z, u(z))\| \right)$$

provided that the Lipschitz constant L is such that $c^*L/\alpha < 1$. Equating the exponents by $N \asymp \sqrt{N_1}$ (i.e., the number of the interpolation points must be proportional to the square root of the number of nodes in the Sinc-quadrature), gives

$$\||Z\|| = \||u - y\|| \leq \frac{c}{\alpha - c^*L} \ln N_1 e^{-c_1\sqrt{N_1}}$$

$$\times \left(\|A^\alpha u_0\| + \int_0^1 \|A^\alpha f(t, u(t))\| dt + \sup_{z \in D_\rho} \|A^\alpha f(z, u(z))\| \right) \qquad (3.287)$$

and due to (3.273), (3.275), (3.280), (3.286), in a stronger norm

$$\||A^\alpha Z\|| = \||A^\alpha (u-y)\|| \le \frac{c}{\beta - \alpha - c^* L} \ln N_1 e^{-c_1 \sqrt{N_1}}$$

$$\times \left(\|A^\beta u_0\| + \int_0^1 \|A^\beta f(t, u(t))\| dt + \sup_{z \in D_\rho} \|A^\beta f(z, u(z))\| \right), \quad (3.288)$$

$$\forall \beta > \alpha > 0, \ \beta - \alpha > c^* L.$$

These estimates show in particular that the operator $\mathcal{A}$ is contractive on G provided that $c^* L / \alpha < 1$ and $N \ge N_0$. Taking into account the Banach fixed point theorem we obtain by usual arguments, that there exists the unique solution of (3.268) in G for which the estimate (3.287) holds.

Thus, we have proven the following result.

Theorem 3.36. *Let A be a densely defined, closed, strongly positive linear operator with the domain $D(A)$ in a Banach space X and the assumptions* (i), (ii), (iii) *hold. Then algorithm* **A1** *defined by* (3.267) *for the numerical solution of the nonlinear problem* (3.246) *possesses a uniform with respect to t exponential convergence rate with estimates* (3.287), (3.288) *provided that $N \asymp \sqrt{N_1}$ and the Lipschitz constant L is sufficiently small.*

Remark 3.37. The same result can be obtained if one uses the interpolation polynomial on the Chebyshev-Gauss-Lobatto grid

$$\omega_N^{CGL} = \{x_{k,N} = x_{k,N}^{CGL} = \cos \frac{(N-j)\pi}{N}, \quad k = 0, 1, \ldots, N, \}$$

where the nodes are zeros of the polynomial $(1 - x^2) T_N'(x)$.

Example 3.38. To have a view of the possible size of the Lipschitz constant let us consider the nonlinear Cauchy problem

$$\frac{d\vec{u}(t)}{dt} + A\vec{u}(t) = \vec{f}(t, \vec{u}(t)), \ t > 0,$$

$$\vec{u}(0) = \vec{u}_0$$

with a linear self-adjoint positive definite operator A such that

$$A = A^* \ge \lambda_0 I, \quad \lambda_0 > 0.$$

In this case, algorithm (3.267) takes the form

$$\vec{y}(t_j) = \vec{y}_j = e^{-At_j} \vec{u}_0 + \sum_{p=1}^N \frac{1}{2} \int_{-1}^{x_{j,N}} e^{-A(x_{j,N} - \eta)/2} L_{p,N-1}(\eta) d\eta \vec{f}(t_p, \vec{y}_p),$$

$$j = 1, \ldots, N.$$

For the error $\vec{z}_j = \vec{y}_j - \vec{u}(t_j) = \vec{y}_j - \vec{u}_j$, we have the equation

$$\vec{z}_j = \sum_{p=1}^{N} \frac{1}{2} \int_{-1}^{x_{j,N}} e^{-A(x_{j,N}-\eta)/2} L_{p,N-1}(\eta) d\eta [\vec{f}(t_p, \vec{y}_p) - \vec{f}(t_p, \vec{u}_p)] + \vec{\psi}_j,$$

$$j = 1, \dots, N,$$

where

$$\vec{\psi}_j = \frac{1}{2} \int_{-1}^{x_{j,N}-\eta} e^{-A(x_{j,N}-\eta)/2} [P_{N-1}(\eta, \vec{u}) - \vec{f}(\frac{\eta+1}{2}, u(\frac{\eta+1}{2}))] d\eta$$

is the truncation error. Using the equation

$$\int_{-1}^{x_{j,N}} e^{-A(x_{j,N}-\eta)/2} L_{p,N-1}(\eta) d\eta$$

$$= -\int_{-1}^{x_{j,N}} e^{-A(x_{j,N}-\eta)/2} \frac{d}{d\eta} \int_{\eta}^{x_{j,N}} L_{p,N-1}(\xi) d\xi d\eta$$

$$= e^{-A(x_{j,N}+1)/2} \int_{-1}^{x_{j,N}} L_{p,N-1}(\eta) d\eta$$

$$+ \frac{1}{2} \int_{-1}^{x_{j,N}} A e^{-A(x_{j,N}-\eta)/2} \int_{\eta}^{x_{j,N}} L_{p,N-1}(\xi) d\xi d\eta,$$

we obtain

$$\vec{z}_j = \sum_{p=1}^{N} \frac{1}{2} \left\{ e^{-A(x_{j,N}+1)/2} \int_{-1}^{x_{j,N}} L_{p,N-1}(\eta) d\eta \right.$$

$$+ \frac{1}{2} \int_{-1}^{x_{j,N}} A e^{-A(x_{j,N}-\eta)/2} \int_{\eta}^{x_{j,N}} L_{p,N-1}(\xi) d\xi d\eta \left. \right\} [\vec{f}(t_p, \vec{y}_p) - \vec{f}(t_p, \vec{u}_p)]$$

$$+ \vec{\psi}_j, \quad j = 1, \dots, N. \tag{3.289}$$

Since A is a self-adjoint, positive definite operator we have

$$\|A e^{-A(x_{j,N}-\eta)/2}\| = \max_{\lambda_0 \le \lambda < \infty} (\lambda e^{-\lambda(x_{j,N}-\eta)/2}) \le \frac{2}{e(x_{j,N}-\eta)},$$

$$\|e^{-A(x_{j,N}+1)/2}\| \le 1.$$

This estimate together with (3.289) and Lemma 3.33 implies

$$\|\vec{z}_j\| \le L \sum_{p=1}^{N} \frac{1}{2} \left| \int_{-1}^{x_{j,N}} L_{p,N-1}(\eta) d\eta \right| \||\vec{z}\|\|$$

$$+ L \frac{1}{2} \sum_{p=1}^{N} \frac{1}{2} \int_{-1}^{x_{j,N}} \|A e^{-A(x_{j,N}-\eta)/2}\| \cdot \left| \int_{-1}^{\eta} L_{p,N}(\xi) d\xi \right| d\eta \cdot \||\vec{z}\|\| + \|\vec{\psi}_j\|$$

$$\leq \frac{1}{2} L \Lambda_j^{(1)}(x_{j,N}) \|\|\vec{z}\|\| + \frac{L}{2e} \int_{-1}^{x_{j,N}} \frac{1}{x_{j,N} - \eta} \Lambda_j^{(2)}(\eta) \|\|\vec{z}\|\| + \|\vec{\psi}_j\|$$

$$\leq \frac{L}{2} \Lambda_j \|\|\vec{z}\|\| + \frac{L\sqrt{\pi}}{2e} \int_{-1}^{x_{j,N}} \frac{1}{\sqrt{x_{j,N} - \eta}} \|\|\vec{z}\|\| + \|\vec{\psi}_j\|$$

$$\leq \left(\sqrt{\frac{\pi}{2}} + \frac{\sqrt{2\pi}}{e} \right) L \|\|\vec{z}\|\| + \|\vec{\psi}_j\|.$$

The last inequality yields the condition

$$L < \frac{\sqrt{2}e}{\sqrt{\pi}e + 2\sqrt{\pi}} \tag{3.290}$$

on the Lipschitz constant L which provides the convergence of the fixed point iteration and a corresponding a priori estimate for $\|\|\vec{z}\|\|$.

Remark 3.39. Given N_1 choose the integer number $N_2 = [\sqrt{N_1}]$ and set $y(t) = P_{N_2-1}(t; y)$ with y defined by algorithm (3.267). To get an error estimate for all $t \in [0,1]$ we represent $Z(t) = u(t) - P_{N_2-1}(2t - 1; y) = u(t) - P_{N_2-1}(2t - 1; u) + [P_{N_2-1}(2t-1; u) - P_{N_2-1}(2t-1; y)]$. Taking into account that the Lebesque constant relating to the Chebyshev interpolation nodes is bounded by $c \ln N_2$ and using the estimates (3.287), $\|P_{N_2-1}(2t - 1; u) - P_{N_2-1}(2t - 1; y)\| \leq c \ln N_2 \|\|Z\|\| \leq c \ln^2 N_2 e^{-c_1 N_2}$ and $\|u(t) - P_{N_2-1}(2t - 1; u)\| \leq c \ln N_2 e^{-c_1 N_2}$ we derive

$$\max_{0 \leq t \leq 1} \|u(t) - P_{N_2-1}(2t - 1; y)\| \leq c \ln^2 N_2 e^{-c_1 N_2}.$$

3.3.4   Modified algorithm for an arbitrary Lipschitz constant

In this section we show how the algorithm above can be modified for a nonlinear case with an arbitrary Lipschitz constant. To this end we suppose that $u(t) \in D(A^\sigma)$, $\sigma > c^* L/2$. We cover the interval $[0,1]$ by the grid $\omega_G = \{t_i = i \cdot \tau : i = 0, 1, \ldots, K, \tau = 1/K\}$ and consider problem (3.246) on each subinterval $[t_{k-1}, t_k]$, $k = 1, \ldots, K$. The substitution $t = t_{k-1}(1 - \xi)/2 + t_k(1 + \xi)/2$, $v(\xi) = u(t_{k-1}(1 - \xi)/2 + t_k(1 + \xi)/2)$ translates the original equation to the differential equation

$$v'(\xi) + \tilde{A}v = \tilde{f}(\xi, v) \tag{3.291}$$

on the reference interval $[-1, 1]$ with $\tilde{A} = \frac{\tau}{2} A$ and with the function $\tilde{f}(\xi, v) = \frac{\tau}{2} f(t_{k-1}(1-\xi)/2 + t_k(1+\xi)/2, u(t_{k-1}(1-\xi)/2 + t_k(1+\xi)/2))$ satisfying the Lipschitz condition with the Lipschitz constant $\tilde{L} = \tau L/2$ which can be made arbitrarily small by the appropriate choice of τ. We cover each subinterval $[t_{k-1}, t_k]$ by the Chebyshev-Gauss-Lobatto grid

$$\omega_{k,N}^{CGL} = \{t_{k,j} : t_{k,j} = t_{k-1}(1 - x_{j,N})/2 + t_k(1 + x_{j,N})/2, j = 0, 1, \ldots, N\},$$

$$x_{j,N} = \cos(\pi(N - j)/N) \tag{3.292}$$

and write $v_k(x_{j,N}) = v_{k,j} = u(t_{k,j})$, $v_{k,0} = v_k$, $u(t_{k,0}) = u(t_k) = u_k$, $\vec{v}_k = [v_{k,j}]_{j=1,\ldots,N}$, $\vec{u}_k = [u(t_{k,j})]_{j=1,\ldots,N}$. Then, algorithm (3.267) with the corresponding Chebyshev-Gauss-Lobatto interpolation polynomial can be applied which provides an exponential accuracy on the subinterval $[t_{k-1}, t_k]$ under the assumption that the initial vector u_{k-1} is known. This is exactly the case for $k = 1$ and, by algorithm (3.267), we obtain a value $v_{1,N} = v_1$ as an approximation for $u(t_1)$. Starting on the subinterval $[t_1, t_2]$ with the approximate initial value v_1 we obtain an approximate solution for this subinterval and so on.

To write down this idea as an algorithm we derive from (3.291) the relation

$$v_{k,j} = e^{-\tilde{A}(1+x_{j,N})} u_{k-1} + \int_{-1}^{x_{j,N}} e^{-\tilde{A}(x_{j,N}-\eta)} \tilde{f}(\eta, v_k(\eta)) d\eta.$$

Denoting by $y_{k,j}$ approximations to $v_{k,j}$, approximating the operator exponential by (3.50) with N_1 nodes and the nonlinearity by the Chebyshev-Gauss-Lobatto interpolation polynomial

$$P_N(\eta, \vec{f}) = \sum_{l=0}^{N} \tilde{f}(x_{l,N}, y_{k,l}) L_{l,N}(\eta),$$

$$L_{l,N}(\eta) = \frac{(1-\eta^2) T'_N(\eta)}{(\eta - x_{l,N}) \frac{d}{d\eta}[(1-\eta^2) T'_N(\eta)]_{\eta=x_{l,N}}},$$

$$\vec{\tilde{f}} = [\tilde{f}(x_{j,N}, y_{k,j})]_{j=0}^{N},$$

we arrive at the following system of nonlinear equations (analogous to (3.267)):

$$y_{k,j} = e_{N_1}^{-\tilde{A}(1+x_{j,N})} y_{k-1} + \int_{-1}^{x_{j,N}} e_{N_1}^{-\tilde{A}(x_{j,N}-\eta)} P_N(\eta, \vec{f}) d\eta, \qquad (3.293)$$

which expresses $y_{k,j}$, $j = 1, 2, \ldots, N$ (in particular $y_{k,N} = y_{k+1}$) through y_{k-1}.

Now, we can formulate the following algorithm.

Algorithm A2. *Given K satisfying (3.296), and N_1 computes the approximate solution of nonlinear problem (3.246) with an arbitrary Lipschitz constant by solving of the nonlinear discrete system (3.293) on each subinterval:*

1. Choose K satisfying (3.296) and N_1 and set $\tau = 1/K$, $t_0 = 0$, $y_0 = u_0$.

2. For $i := 1$ step 1 to K do

   2.1. Set $t_i = t_{i-1-1-3} + \tau$ and find the approximate solution $y_{i,j}$, $j = 1, 2, \ldots, N$ of problem (3.246) on the Chebyshev-Gauss-Lobatto grid (3.292) covering the interval $[t_{i-1-1-3}, t_i]$ by algorithm (3.293) using $y_{i-1-1-3}$ as the initial value.

   2.2. Set $y_i = y_{i,N}$.

Now, let us analyze the error $z_{k,j} = u(t_{k,j} - y_{k,j}) = v_{k,j} - y_{k,j}$ of this algorithm. We have the representation

$$z_{k,j} = \psi_{k,j} + \sum_{p=0}^{3} \psi_{k,j}^{(p)}, \tag{3.294}$$

where

$$\psi_{k,j} = e_{N_1}^{-\tilde{A}(x_{j,N}+1)} z_{k-1},$$

$$\psi_{k,j}^{(0)} = [e^{-\tilde{A}(x_{j,N}+1)} - e_{N_1}^{-\tilde{A}(x_{j,N}+1)}] u_{k-1},$$

$$\psi_{k,j}^{(1)} = \int_{-1}^{x_{j,N}} [e^{-\tilde{A}(x_{j,N}-\eta)} - e_{N_1}^{-\tilde{A}(x_{j,N}-\eta)}] \tilde{f}(\eta, v_k(\eta)) d\eta,$$

$$\psi_{k,j}^{(2)} = \int_{-1}^{x_{j,N}} e_{N_1}^{-\tilde{A}(x_{j,N}-\eta)} [\tilde{f}(\eta, v_k(\eta)) - \sum_{l=0}^{N} \tilde{f}(x_{l,N}, v_{k,l}) L_{l,N}(\eta)] d\eta,$$

$$\psi_{k,j}^{(3)} = \int_{-1}^{x_{j,N}} e_{N_1}^{-\tilde{A}(x_{j,N}-\eta)} \sum_{l=0}^{N} [\tilde{f}(x_{l,N}, v_{k,l}) - \tilde{f}(x_{l,N}, y_{k,l})] L_{l,N}(\eta) d\eta.$$

Under the same assumptions and analogously to (3.272), (3.274), (3.279), (3.285) we obtain the following estimates:

$$\|\psi_{k,j}\| \leq \frac{c}{\alpha_k} \|\tilde{A}^{\alpha_k} z_{k-1}\|,$$

$$\|\psi_{k,j}^{(0)}\| \leq \frac{c}{\alpha_k} \exp\left(-\sqrt{\frac{\pi d \alpha_k}{2}} (N_1 + 1)\right) \|\tilde{A}^{\alpha_k} u_{k-1}\|,$$

$$\|\psi_{k,j}^{(1)}\| \leq \frac{c}{\alpha_k} \exp\left(-\sqrt{\frac{\pi d \alpha_k}{2}} (N_1 + 1)\right) \int_{-1}^{1} \|\tilde{A}^{\alpha_k} \tilde{f}(t, v_k(t))\| dt, \tag{3.295}$$

$$\|\psi_{k,j}^{(2)}\| \leq \frac{c}{\alpha_k} \ln N \rho^{-N} \sup_{z \in D_{\rho_k}} \|\tilde{A}^{\alpha_k} \tilde{f}(z, v_k(z))\|,$$

$$\|\psi_{k,j}^{(3)}\| \leq \frac{c^*}{\alpha_k} \frac{L\tau}{2} \||\vec{z}_k\||,$$

where α_k are some positive numbers, D_{ρ_k} are the analyticity ellipses for $\tilde{A}^{\alpha_k} \tilde{f}(z, v_k(z))$ and $\||\vec{z}_k\|| = \max_{1 \leq j \leq N} \|z_{k,j}\|$. Choosing $\tau = 1/K$ such that

$$\frac{c^*}{\alpha_k} \frac{L\tau}{2} < 1, \tag{3.296}$$

we obtain from (3.294), (3.295)

$$\||\vec{z}_k\|| = \max_{1 \leq j \leq N} \|z_{k,j}\| \leq \frac{c(\tau/2)^{\alpha_k}}{\alpha_k - c^* L\tau/2} \left\{ \|A^{\alpha_k} z_{k-1}\| \right.$$

$$+ \left[\int_{t_{k-1}}^{t_k} \|A^{\alpha_k} f(t, u(t))\| dt + \|A^{\alpha_k} u(t_{k-1})\| \right] \exp\left(-\sqrt{\frac{\pi d \alpha_k}{2}} (N_1 + 1) \right)$$

$$+ \frac{\tau}{2} \ln N \rho^{-N} \sup_{z \in D_{\rho_k}} \|A^{\alpha_k} f(t_k(z), u(t_k(z)))\| \bigg\},$$

$$k = 2, 3, \ldots, K.$$

$$t_k(z) = t_{k-1} \frac{1-z}{2} + t_k \frac{1+z}{2}. \tag{3.297}$$

Equalizing the exponents by setting $N \asymp \sqrt{N_1}$ (i.e., the number of the interpolation points on each subinterval must be proportional to the square root of the number of nodes in the Sinc approximation of the operator exponential) we obtain from (3.297)

$$\||A^{\alpha_{k+1}} \vec{z}_k\|| \le \frac{c(\tau/2)^{\alpha_k}}{\alpha_k - \alpha_{k+1} - c^* L \tau/2} \bigg\{ \||A^{\alpha_k} z_{k-1}\||$$

$$+ \ln N_1 e^{-c_1 \sqrt{N_1}} \left[\int_{t_{k-1}}^{t_k} \|A^{\alpha_k} f(t, u(t))\| dt + \|A^{\alpha_k} u(t_{k-1})\| \right. \tag{3.298}$$

$$+ \frac{\tau}{2} \sup_{z \in D_{\rho_k}} \|A^{\alpha_k} f(t_k(z), u(t_k(z)))\| \bigg] \bigg\},$$

$$k = 2, 3, \ldots, K,$$

where α_k satisfy

$$\alpha_k - \alpha_{k+1} - c^* L \tau/2 > 0,$$
$$0 < \alpha_k \le \sigma, \quad k = 1, 2, \ldots, K. \tag{3.299}$$

Taking into account that $z_0 = 0$ we have the estimate

$$\||A^{\alpha_2} \vec{z}_k\|| \le \frac{c(\tau/2)^{\alpha_1}}{\alpha_1 - \alpha_2 - c^* L \tau/2} \ln N_1 e^{-c_1 \sqrt{N_1}} \left[\|A^{\alpha_1} u_0\| + \int_{t_0}^{t_1} \|A^{\alpha_1} f(t, u(t))\| dt \right.$$

$$+ \frac{\tau}{2} \sup_{z \in D_{\rho_1}} \|A^{\alpha_k} f(t_k(z), u(t_k(z)))\| \bigg]$$

for $k = 1$. Estimate (3.298) can be rewritten in the form

$$w_k \le \mu_k (g_k + w_{k-1}), \quad k = 1, 2, \ldots, K$$

with

$$w_k = \||A^{\alpha_2} \vec{z}_k\||, \quad \mu_k = \frac{c(\tau/2)^{\alpha_1}}{\alpha_k - \alpha_{k+1} - c^* L \tau/2},$$

$$g_k = \ln N_1 e^{-c_1 \sqrt{N_1}} \left[\int_{t_{k-1}}^{t_k} \|A^{\alpha_k} f(t, u(t))\| dt + \|A^{\alpha_k} u(t_{k-1})\| \right.$$

$$+ \frac{\tau}{2} \sup_{z \in D_{\rho_k}} \|A^{\alpha_k} f(t_k(z), u(t_k(z)))\| \bigg]$$

which yields

$$w_k \le \mu_k(g_k + w_{k-1}), \quad k = 1, 2, \dots, K$$

and further recursively

$$w_k \le \mu_k g_k + \mu_{k-1}\mu_k g_{k-1} + \cdots + \mu_1\mu_2 \cdots \mu_k g_1. \tag{3.300}$$

Conditions (3.299) imply

$$0 < \alpha_{k+1} < \alpha_1 - c^* L k \tau / 2 > 0, \quad k = 1, 2, \dots, K.$$

Let us choose $\alpha_1 = \sigma$, $\varepsilon \in (0, \frac{\sigma - c^* L/2}{c^* L})$ and

$$\alpha_{k+1} = \sigma - \left(\frac{1}{2} + \varepsilon\right) c^* L k \tau, \quad k = 1, 2, \dots, K.$$

Then we have

$$\rho_k = \frac{c(\tau/2)^{\alpha_k}}{\varepsilon c^* L \tau} = \frac{c(\tau/2)^{\alpha_k - 1}}{2\varepsilon c^* L} < \frac{c}{2\varepsilon c^* L}(\tau/2)^{\sigma - (0.5 + \varepsilon)c^* L} = q$$

and (3.300) implies

$$\max_{1 \le k \le K} w_k \le \max\{q^K, q\} \sum_{p=1}^{K} g_p$$

or

$$\max_{1 \le k \le K} |||A^{\alpha_2}\vec{z}_k||| \le \max\{q^K, q\} \ln N_1 e^{-c_1\sqrt{N_1}} \sum_{k=1}^{K} \left[\int_{t_{k-1}}^{t_k} \|A^{\alpha_k} f(t, u(t))\| dt \right.$$

$$\left. + \|A^{\alpha_k} u(t_{k-1})\| + \frac{\tau}{2} \sup_{z \in D_{\rho_k}} \|A^{\alpha_k} f(t_k(z), u(t_k(z)))\| \right]. \tag{3.301}$$

Thus, we have proven the following second main result on the rate of convergence of the algorithm **A2**.

Theorem 3.40. *Let A be a densely defined closed strongly positive linear operator with the domain $D(A)$ in a Banach space X and the assumptions* **(i)**, **(ii)**, **(iii)** *hold. If the solution of the nonlinear problem (3.246) belongs to the domain $D(A^\sigma)$ with $\sigma > c^* L/2$, then algorithm* **A2** *possesses a uniform with respect to t exponential convergence rate with estimate (3.301), provided that $N \asymp \sqrt{N_1}$ and the chosen number of subintervals K satisfies (3.296).*

3.3.5 Implementation of the algorithm

Algorithm (3.267) represents a nonlinear system of algebraic equations which can be solved by the fixed point iteration

$$
\begin{aligned}
y_j^{(m+1)} = {} & \frac{h}{2\pi i} \sum_{k=-N_1}^{N_1} \mathcal{F}_h(x_{j,N}, kh) \\
& + \frac{h}{4\pi i} \int_{-1}^{x_j} \sum_{k=-N_1}^{N_1} F_A(\xi, x_j - \eta) P_{N-1}(\eta; f(\cdot, y^{(m)})) d\eta,
\end{aligned}
\tag{3.302}
$$

$$
j = 1, \ldots, N, \quad m = 0, 1, \ldots.
$$

Since the operator $\mathcal{A}$ is contractive we obtain the inequality

$$
\||y_j^{(m+1)} - y_j^{(m)}\|| \leq Lc^* \||y_j^{(m)} - y_j^{(m-1)}\||
$$

which justifies the convergence of the fixed point iteration (3.302) with the speed of a geometric progression with the denominator $Lc^* < 1$, provided the assumptions of Theorem 3.36 hold.

Let us estimate the asymptotical computational costs of our method and a possible alternative polynomially convergent method (e.g., step-by-step implicit Euler method) to arrive at a given tolerance ε. Assuming the time step τ and the spatial step h in the Euler scheme to be equal, we have asymptotically to make $\frac{t^*}{\varepsilon}$ steps in order to arrive at a tolerance ε at a given fixed point $t = t^*$. At each step the nonlinear equation $\tau f(t_{k+1}, y_{k+1}) - (\tau A y_{k+1} + I) = y_k$ should be solved, where y_k is an approximation for $u(t_k)$. Assuming the computational costs for the solution of this nonlinear equation to be M, we arrive at the total computational costs for the Euler method $T_E \asymp t^* M / \varepsilon$. From the asymptotical relation $\ln N_1 e^{-c\sqrt{N_1}} < e^{-c_1 \sqrt{N_1}} \asymp \varepsilon$ we obtain that in our algorithm $N \asymp \sqrt{N_1} \asymp \ln(1/\varepsilon)$. It is natural to assume that the computational costs for the numerical solution of the nonlinear equation (3.302) (or (3.268)) are not greater than $NM \asymp \sqrt{N_1} M$. Then the total costs of our algorithm are $T_O \asymp M \ln(1/\varepsilon) \ll T_E$ for ε small enough.

Example 3.41. Let us consider the nonlinear initial value problem

$$
\begin{aligned}
& u'(t) + u(t) = \mu e^{-2t} - \mu [u(t)]^2, \quad t \in (-1, 1], \\
& u(-1) = e
\end{aligned}
\tag{3.303}
$$

with the exact solution $u(t) = e^{-t}$ (independent of μ). The equivalent Volterra integral equation is

$$
u(t) = \varphi(t) - \mu \int_{-1}^{t} e^{-(t-s)} u^2(s) ds
$$

where

$$
\varphi(t) = e^{-t} + \mu[e^{1-t} - e^{-2t}].
$$

Algorithm (3.267) combined with the fixed point iteration takes in this case the form

$$y_j^{(m+1)} = \varphi_j - \mu \sum_{p=1}^{N} \alpha_{p,j} [y_p^{(m)}]^2,$$

$$y_j^{(0)} = 1/2, \ j = 1, \ldots, N, \ m = 0, 1, \ldots$$

where

$$\alpha_{p,j} = e^{-x_{j,N}} \int_{-1}^{x_{j,N}} L_{p,N-1}(s) e^s ds,$$

$$y_j = y(x_{j,N}), \ \varphi_j = \varphi(x_{j,N}).$$

(3.304)

The algorithm was implemented in Maple 8 (Digits=30) for $\mu = 1/4$ where integrals (3.304) were computed analytically. We denote by It the number of iterations necessary to satisfy the interruption criterium $|y_{J,N}^{(m+1)} - y_{J,N}^{(m)}| < e^{-N} \cdot 10^{-2}$ and accept $y_j^{(It)} = y_{J,N}^{(m+1)}$ as the approximate solution. The error is computed as $\varepsilon_N = \|u - y\|_{N,\infty} = \max_{1 \leq j \leq N} |u(x_{j,N}) - y_j^{(It)}|$. The numerical results are given by Table 3.5 and confirm our theory.

N	ε_N	It
2	0.129406	6
4	0.626486 e-2	8
8	0.181353 e-5	9
16	0.162597 e-14	16
32	0.110000 e-28	26

Table 3.5: The error of algorithm (3.267) for problem (3.303).

Example 3.42. Let us consider the problem

$$\frac{\partial u}{\partial t} + Au = f(t, u(t)),$$

$$u(-1) = u_0$$

with the linear operator A given by

$$D(A) = \{w(x) \in H^2(0,1) : \ w'(0) = 0, w'(1) = 0\},$$
$$Av = -w'' \quad \forall w \in D(A),$$

with the nonlinear operator f given by

$$f(t, u) = -2tu^2$$

and with the initial condition given by

$$u_0 = u(-1, x) = 1/2.$$

Since the numerical algorithm supposes that the operator coefficient is strongly positive, we shift its spectrum by the variables transform $u(t, x) = e^{d^2 t} v(t, x)$ with a real number d. Then we obtain the problem

$$\frac{\partial v}{\partial t} + A_d v = f_d(t, v(t)),$$

$$v(-1) = v_0$$

with the linear operator A_d given by

$$D(A_d) = D(A),$$

$$A_d w = A w + d^2 w \quad \forall w \in D(A_d),$$

with the nonlinear operator

$$f_d(t, v) = -2t e^{d^2 t} v^2,$$

and with the initial condition

$$v_0 = v(-1, x) = e^{d^2}/2.$$

It is easy to check that the exact solution of this problem is

$$v(t, x) = e^{-d^2 t}/(1 + t^2).$$

The equivalent Volterra integral equation for v has the form

$$v(t, x) = \frac{1}{2} e^{-A_d(t+1)} e^{d^2} - 2 \int_{-1}^{t} e^{-A_d(t-s)} s e^{d^2 s} [v(s, \cdot)]^2 ds.$$

Returning to the unknown function u, the integral equation takes the form

$$u(t, x) = \frac{1}{2} e^{-A_d(t+1)} e^{d^2(t+1)} - 2 \int_{-1}^{t} e^{-A_d(t-s)} s e^{-d^2 s} [u(s, \cdot)]^2 ds.$$

Our algorithm was implemented in Maple with numerical results given by Table 3.6 where $\varepsilon_N = \max_{1 \leq j \leq N} \varepsilon_{j,N}$, $\varepsilon_{j,k,N} = |u(x_{j,N}, kh) - y_{j,k}|$, $j = 1, \ldots, N$, $k = -N_1, \ldots, N_1$. The numerical results are in a good agreement with Theorem 3.36.

Example 3.43. This example deals with the two-dimensional nonlinear problem

$$\frac{\partial u}{\partial t} + A u = f(t, u(t)),$$

$$u(0) = u_0,$$

(3.305)

N	ε_N	It
4	0.8 e-1	12
8	0.7 e-3	10
16	0.5 e-6	11
32	0.3 e-12	12

Table 3.6: The error of algorithm (3.267) for problem (3.303).

where

$$D(A) = \{w(x,y) \in H^2(\Omega) : w|_{\partial\Omega} = 0\},$$
$$Av = -\Delta v \; \forall v \in D(A), \tag{3.306}$$
$$\Omega = [0,1] \times [0,1]$$

with the nonlinear operator f given by

$$f(t,u) = -u^3 + e^{-6\pi^2 t} \sin^3 \pi x \sin^3 \pi y \tag{3.307}$$

and with the initial condition given by

$$u_0 = u(0,x,y) = \sin \pi x \sin \pi y. \tag{3.308}$$

The exact solution is given by $u = e^{-2\pi^2 t} \sin \pi x \sin \pi y$. Algorithm (3.267) with $N = \sqrt{N_1}$ Chebyshev-Gauss-Lobatto nodes combined with the fixed point iteration provides the error which is presented in Table 3.7.

N	ε_N	It
4	.3413e-6	12
8	.1761e-6	10
16	.8846e-7	14
32	.5441e-8	14

Table 3.7: The error ε_N of algorithm (3.267) for problem (3.305)- (3.308).

Example 3.44. Let us consider again the nonlinear initial value problem (3.303) and apply the algorithm **A2** for various values of the Lipschitz constant 2μ. Inequality (3.290) guarantees the convergence of algorithm (3.267) combined with the fixed point iteration for $\mu < 0.4596747673$. Numerical experiments indicate the convergence for $\mu > 0.4596747673$ but beginning with $\mu \approx 1$, the process becomes divergent and algorithm **A2** should be applied. The corresponding results for various μ are presented in Table 3.8.

Here the degree of the interpolation polynomial is $N = 16$, K is the number of subintervals of the whole interval $[-1;1]$, It denotes the number of the iterations in order to arrive at the accuracy $\exp(-N) * 0.01$.

μ	K	It
0.9	1	22
1	2	20
10	32	20
20	50	25
50	128	25
100	256	24

Table 3.8: The results of algorithm **A2** for problem (3.303) with various values of the Lipschitz constant μ.

3.4 Exponentially convergent Duhamel-Like algorithms for differential equations with an operator coefficient possessing a variable domain in a Banach space

3.4.1 Introduction

This section deals with a special class of parabolic partial differential equations with time-dependent boundary conditions in an abstract setting, which are associated with the first-order differential equation in a Banach space X,

$$\frac{du(t)}{dt} + A(t)u(t) = f(t), \quad u(0) = u_0. \tag{3.309}$$

Here t is a real variable, the unknown function $u(t)$, and the given function $f(t)$ take values in X, and $A(t)$ is a given function whose values are densely defined, closed linear operators in X with domains $D(A, t)$ depending on the parameter t. In some special cases, it is possible to translate problem (3.309) to an operator differential equation, in which operator coefficient $\tilde{A}$ possesses domain $D(\tilde{A})$ independent of t. For example, let us consider the problem

$$\frac{\partial u(x,t)}{\partial t} = \frac{\partial^2 u}{\partial x^2} + f(x,t),$$
$$u_x(0,t) - \alpha(t)u(0,t) = \phi_1(t), \quad u_x(1,t) + \beta(t)u(1,t) = \phi_2(t), \tag{3.310}$$
$$u(x,0) = u_0(x),$$

where f, α, β, ϕ_1, ϕ_2, and u_0 are given smooth functions. This problem belongs to class (3.309). The operator coefficient A is as follows:

$$D(A,t) = \{u(x) \in H^2(0,1) : u'(0) - \alpha(t)u(0) = \phi_1(t), \ u'(1) + \beta(t)u(1) = \phi_2(t)\},$$
$$Au = -u''(x) \quad \forall u \in D(A,t).$$

Substitution of (compare with [33])

$$u(x,t) = e^{r(x,t)}v(x,t) \tag{3.311}$$

with function $r(x,t)$ satisfying

$$r_x(0,t) + \alpha(t) = 0, \quad r_x(1,t) - \beta(t) = 0 \tag{3.312}$$

(one can, for example, choose $r(x,t) = -\alpha(t) \cdot x + \frac{1}{2}[\alpha(t) + \beta(t)] \cdot x^2$) transforms (3.310) into

$$\begin{aligned}
&\frac{\partial v(x,t)}{\partial t} = \frac{\partial^2 v}{\partial x^2} + 2r_x(x,t)\frac{\partial v}{\partial x} + (r_{xx} - r_t)v + e^{-r(x,t)}f(x,t), \\
&v_x(0,t) = e^{-r(0,t)}\phi_1(t), \quad v_x(1,t) = e^{-r(0,t)}\phi_2(t), \\
&v(x,0) = e^{-r(x,0)}u_0(x).
\end{aligned} \tag{3.313}$$

The new operator coefficient $\tilde{A}(t)$ is defined by

$$\begin{aligned}
&D(\tilde{A},t) = \{v(x) \in H^2(0,1): \ v'(0) = e^{-r(0,t)}\phi_1(t), \ v'(1) = e^{-r(1,t)}\phi_2(t)\}, \\
&-\tilde{A}(t)u = v''(x) + 2r_x(x,t)v' + (r_{xx} - r_t)v \quad \forall v \in D(\tilde{A},t).
\end{aligned} \tag{3.314}$$

The operator coefficient $\tilde{A}$ becomes more complicated and is not self-adjoint anymore. However, in the homogeneous case $\phi_1(t) \equiv 0$, $\phi_2(t) \equiv 0$ due to substitution (3.311), we obtain a bounded operator $R(t)$ as the multiplicative operator $R(t) = e^{r(x,t)}$ such that the operator $\tilde{A}(t) = [R(t)]^{-1}AR(t)$ possesses the domain independent of t. Under this and some other assumptions it was shown in [33, 31] that the initial value problem (3.310) has a unique solution. Unfortunately there is not a constructive way to find such an operator $R(t)$ in the general case.

The variable domain of an operator can be described by a separate equation, and then we have an abstract problem of the kind

$$\begin{aligned}
&\frac{du(t)}{dt} = A(t)u(t), \quad 0 \leq s \leq t \leq T, \\
&L(t)u(t) = \Phi(t)u(t) + f(t), \quad 0 \leq s \leq t \leq T, \\
&u(s) = u_0
\end{aligned} \tag{3.315}$$

instead of (3.309). Here $L(t)$ and $\Phi(t)$ are appropriate linear operators defined on the boundary of the spatial domain, and the second equation represents an abstract model of the time-dependent boundary condition. An existence and uniqueness result for this problem was proven in [13]. Incorporating the boundary condition into the definition of the operator coefficient of the first equation, one obtains problem (3.309).

The literature concerning discretizations of such problems in an abstract setting is rather limited (see, e.g., [53], where the Euler difference approximation of the first accuracy order for problem (3.309) with the time-dependent domain was considered, and the references therein). The classical Duhamel integral together with discretization of high accuracy order for the two-dimensional heat equation with the time-dependent inhomogeneous Dirichlet boundary condition was proposed in [28]. It is clear that the discretization (with respect to t) is more complicated than in the case of a t-independent domain $D(A)$, since the inclusion $y_k = y(t_k) \in D(A, t_k)$ of the approximate solution y_k at each discretization point t_k should be additionally checked and guaranteed.

In this section, we consider the problem

$$\frac{du(t)}{dt} + A(t)u(t) = f(t),$$
$$\partial_1 u(t) + \partial_0(t)u(t) = g(t), \tag{3.316}$$
$$u(0) = u_0,$$

where $u(t)$ is the unknown function $u : (0, T) \to D(A) \subset X$ with values in a Banach space X, $f(t)$ is a given measurable function $f : (0, T) \to X$ from $L_q(0, T; X)$ with the norm $\|f\| = \{\int_0^T \|f\|_X^q dt\}^{1/q}$, $A(t) : D(A) \in X$ is a densely defined, closed linear operator in X with a time-independent domain $D(A)$, $g : (0, T) \to Y$ is a given function from $L_q(0, T; Y)$ with values in some other Banach space Y, and $\partial_1 : D(A) \to Y$ (independent of t!), $\partial_0(t) : D(A) \to Y$ (can depend on t!) are linear operators. In applications, the second equation above is just the time-dependent boundary condition with appropriate operators ∂_1, ∂_0 acting on the boundary of the spatial domain (see section 3.4.6 for examples). For this reason, we call this equation an abstract (time-dependent) boundary condition.

Incorporating the boundary condition into the definition of the operator coefficient in the first equation, we get a problem of the type (3.309) with a variable domain. The difficulties of the discretization were mentioned above and were well discussed in [53]. The separation of this condition in (3.316) will allow us below to use an abstract Duhamel-like technique in order to reduce the problem to another one including operators with t-independent domains.

3.4.2 Duhamel-like technique for first-order differential equations in Banach space

In this subsection we consider a particular case of problem (3.316),

$$\frac{du(t)}{dt} + Au(t) = f(t),$$
$$\partial_1 u(t) + \partial_0(t)u(t) = g(t), \quad u(0) = u_0, \tag{3.317}$$

where the operator A and its domain $D(A)$ in some Banach space X are independent of t, $f(t)$ is a given measurable function $f : (0, T) \to X$ from the space $L_q(0, T; X)$ with the norm $\|f\| = \{\int_0^T \|f\|_X^q dt\}^{1/q}$, $g : (0, T) \to Y$ is a given function from $L_q(0, T; Y)$ with values in some other Banach space Y, and $\partial_1 : D(A) \to Y$ (independent of t!), $\partial_0(t) : D(A) \to Y$ (can depend on t!) are linear operators.

We represent the solution in the form

$$u(t) = w(t) + v(t), \tag{3.318}$$

with v and w satisfying

$$\frac{dv}{dt} + Av = f(t),$$
$$\partial_1 v = 0, \quad v(0) = u_0 \tag{3.319}$$

and

$$\frac{dw(t)}{dt} + Aw = 0,$$
$$\partial_1 w(t) = -\partial_0(t)u(t) + g(t), \tag{3.320}$$
$$w(0) = 0,$$

respectively. Introducing the operator $A^{(1)} : D(A^{(1)}) \to X$ independent of t and defined by

$$D(A^{(1)}) = \{u \in D(A) : \partial_1 u = 0\},$$
$$A^{(1)}u = Au \quad \forall u \in D(A^{(1)}), \tag{3.321}$$

we can rewrite problem (3.319) in the form

$$\frac{dv}{dt} + A^{(1)}v = f(t),$$
$$v(0) = u_0. \tag{3.322}$$

We suppose that the operator A is such that the operator $A^{(1)}$ is strongly positive (see 2 or [4, 20, 27]) (m-sectorial in the sense of [14]); i.e., there exist a positive constant M_R and a fixed $\theta \in [0, \pi/2)$ such that on the rays from the origin building a sector $\Sigma_\theta = \{z \in \mathbb{C} : |\arg(z)| \le \theta\}$ and outside this sector, the following resolvent estimate holds:

$$\|(zI - A^{(1)})^{-1}\| \le \frac{M_R}{1 + |z|}. \tag{3.323}$$

Then under some assumptions with respect to $f(t)$, the solution of (3.322) can be represented by (3.4), (3.5) (see [54])

$$v(t) = e^{-A^{(1)}t}u_0 + \int_0^t e^{-A^{(1)}(t-\tau)}f(\tau)d\tau. \tag{3.324}$$

Note that problem (3.322) with a strongly positive operator $A^{(1)}$ can also be solved numerically by the exponentially convergent algorithm from subsection 3.1.4.

Returning to problem (3.320) we introduce the auxiliary function $W(\lambda, t)$ by

$$
\begin{aligned}
\frac{\partial W(\lambda, t)}{\partial t} + AW(\lambda, t) &= 0, \\
\partial_1 W(\lambda, t) &= -\partial_0(\lambda)u(\lambda) + g(\lambda), \\
W(\lambda, 0) &= 0,
\end{aligned}
\tag{3.325}
$$

where the abstract boundary condition is now independent of t. The next result gives a representation of the solution of problem (3.320) through the function W by the so-called Duhamel integral.

Theorem 3.45. *The solution of problem (3.320) with a t-dependent boundary condition can be represented through the solution of problem (3.325) with a t-independent boundary condition by the following Duhamel integral:*

$$
w(t) = \frac{d}{dt} \int_0^t W(\lambda, t - \lambda) d\lambda = \int_0^t \frac{\partial}{\partial t} W(\lambda, t - \lambda) d\lambda.
\tag{3.326}
$$

Proof. Let us show that the function (3.326) in fact satisfies (3.320) (compare with the classical representation by Duhamel's integral [28, 55]).

Actually, the initial condition $w(0) = 0$ obviously holds true. Due to the first representation in (3.326) and (3.325), we have

$$
\begin{aligned}
\partial_1 w(t) &= \frac{d}{dt} \int_0^t \partial_1 W(\lambda, t - \lambda) d\lambda = \frac{d}{dt} \int_0^t [-\partial_0(\lambda)u(\lambda) + g(\lambda)] d\lambda \\
&= -\partial_0(t)u(t) + g(t);
\end{aligned}
\tag{3.327}
$$

i.e., the boundary condition $\partial_1 w(t) = -\partial_0(t)u(t) + g(t)$ is also fulfilled. Due to the second representation in (3.326) we obtain

$$
\begin{aligned}
\frac{dw(t)}{dt} + Aw(t) &= \frac{d}{dt} \int_0^t \frac{\partial}{\partial t} W(\lambda, t - \lambda) d\lambda + Aw(t) \\
&= \left. \frac{\partial}{\partial t} W(\lambda, t - \lambda) \right|_{\lambda=t} + \int_0^t \frac{\partial^2}{\partial t^2} W(\lambda, t - \lambda) d\lambda + \int_0^t \frac{\partial}{\partial t} AW(\lambda, t - \lambda) d\lambda \\
&= -AW(\lambda, t - \lambda)|_{\lambda=t} + \int_0^t \frac{\partial}{\partial t} \left[\frac{\partial}{\partial t} W(\lambda, t - \lambda) + AW(\lambda, t - \lambda) \right] d\lambda = 0;
\end{aligned}
\tag{3.328}
$$

i.e., the first equation in (3.320) holds true, which completes the proof. $\square$

To make the boundary condition in (3.325) homogeneous and independent of t, we introduce the operator $B : Y \to D(A)$ by

$$
\begin{aligned}
A(By) &= 0, \\
\partial_1 By &= y;
\end{aligned}
\tag{3.329}
$$

i.e., $\partial_1 B$ is a projector on Y.

Remark 3.46. To show the existence of the operator B, we set $By = v$. Then the problem (3.329) is equivalent to the problem

$$Av = 0,$$
$$\partial_1 v = y.$$

We assume that there exists an operator $E : Y \to D(A)$ such that $\partial_1 Ey = y$ (compare with $E(t)$ from [53]). The existence of this operator for an elliptic operator A is proven in [67, section 5.4, Theorem 5.4.5]. Let us introduce $v_1 = v - Ey \in D(A)$, and then

$$Av_1 = -AEy,$$
$$\partial_1 v_1 = 0.$$

Taking into account (3.321) we have the problem

$$A^{(1)} v_1 = -AEy,$$

which has a solution since operator $A^{(1)}$ is strongly positive. Thus, we have

$$B = (I - [A^{(1)}]^{-1} A) E.$$

Given operator B we change the dependent variable by

$$W(\lambda, t) = B[-\partial_0(\lambda) u(\lambda) + g(\lambda)] + W_1(\lambda, t). \tag{3.330}$$

For the new dependent variable we have the problem

$$\frac{\partial W_1(\lambda, t)}{\partial t} + A W_1(\lambda, t) = 0,$$
$$\partial_1 W_1(\lambda, t) = 0, \tag{3.331}$$
$$W_1(\lambda, 0) = -B[-\partial_0(\lambda) u(\lambda) + g(\lambda)]$$

or, equivalently,

$$\frac{\partial W_1(\lambda, t)}{\partial t} + A^{(1)} W_1(\lambda, t) = 0,$$
$$W_1(\lambda, 0) = -B[-\partial_0(\lambda) u(\lambda) + g(\lambda)]. \tag{3.332}$$

Using the operator exponential, the solution of this problem can be given in the form

$$W_1(\lambda, t) = -e^{-A^{(1)} t} B[-\partial_0(\lambda) u(\lambda) + g(\lambda)]. \tag{3.333}$$

Now, taking into account the substitution (3.330) and the last representation, we obtain

$$W(\lambda, t) = -[e^{-A^{(1)} t} - I] B[-\partial_0(\lambda) u(\lambda) + g(\lambda)], \tag{3.334}$$

where $I : X \to X$ is the identity operator. Due to relations (3.318), (3.326), and (3.334), we arrive at the representation

$$u(t) = v(t) + \frac{\partial}{\partial t} \int_0^t W(\lambda, t - \lambda) d\lambda$$

$$= v(t) - \int_0^t \frac{\partial}{\partial t} \left\{ e^{-A^{(1)}(t-\lambda)} B[-\partial_0(\lambda) u(\lambda) + g(\lambda)] \right\} d\lambda \tag{3.335}$$

and further the following boundary integral equation:

$$\partial_0(t) u(t) = \partial_0(t) v(t) - \partial_0(t) \int_0^t \frac{\partial}{\partial t} \left\{ e^{-A^{(1)}(t-\lambda)} B[-\partial_0(\lambda) u(\lambda) + g(\lambda)] \right\} d\lambda. \tag{3.336}$$

The last equation can also be written in the form

$$\partial_0(t) u(t) = \partial_0(t) v(t) + \partial_0(t) \int_0^t A^{(1)} e^{-A^{(1)}(t-\lambda)} B[-\partial_0(\lambda) u(\lambda) + g(\lambda)] d\lambda. \tag{3.337}$$

After determining $\partial_0(t) u(t)$ from (3.337), we can find the solution of problem (3.317) by (3.324), (3.335) using the exponentially convergent algorithms for the operator exponential from the subsection 3.1.4 (see also [27, 18]).

Let us introduce the operator-valued kernel $K(t - \tau) = A^{(1)} e^{-A^{(1)}(t-\tau)} B$, the functions $\theta(t) = \partial_0(t) u(t)$ and $F(t) = \partial_0(t) v(t) + \partial_0(t) \int_0^t A^{(1)} e^{-A^{(1)}(t-\lambda)} B g(\lambda) d\lambda$, and the operator $V : L_q(0, T; D(A)) \to L_q(0, T; Y)$ defined by

$$V(t) y(\cdot) = \partial_0(t) \int_0^t K(t - \lambda) y(\lambda) d\lambda. \tag{3.338}$$

Then (3.336) can be written in the fixed point iteration form

$$\theta(t) = V(t) \theta(\cdot) + F(t). \tag{3.339}$$

3.4.3 Existence and uniqueness of the solution of the integral equation

To prove the existence and uniqueness result for the equivalent equations (3.336), (3.337), and (3.339), we make the following hypotheses (see subsection 3.4.6 for examples):

(A1) There exists a positive constant c such that

$$\|\partial_0(t)\|_{X \to Y} \le c \quad \forall t \in [0, T] \tag{3.340}$$

(compare with H1 from [53]).

(A2) For some $p \geq 1$ it holds that

$$\left[\int_0^t \| A^{(1)} e^{-A^{(1)}(t-\lambda)} B \|_{Y \to X}^p d\lambda \right]^{1/p} \leq c \quad \forall t \in [0, T]. \tag{3.341}$$

(A3) The function

$$F(t) = \partial_0(t)v(t) - \partial_0(t) \int_0^t \frac{\partial}{\partial t} e^{-A^{(1)}(t-\lambda)} Bg(\lambda) d\lambda$$

$$= \partial_0(t)v(t) + \partial_0(t) \int_0^t A^{(1)} e^{-A^{(1)}(t-\lambda)} Bg(\lambda) d\lambda : (0, T) \to Y \tag{3.342}$$

belongs to the Banach space $L_q(0, T; Y)$ and

$$\| F \|_{L_q(0,T;Y)} = \left\{ \int_0^T \| F(\tau) \|_Y^q d\tau \right\}^{1/q} \leq c, \tag{3.343}$$

where q is such that $\frac{1}{p} + \frac{1}{q} = 1$ for p defined in (A2).

Let us define the sequences

$$\theta_j(t) = V(t)\theta_{j-1}(\cdot) + F(t),$$
$$\delta_j(t) = \theta_j(t) - \theta^*(t), \tag{3.344}$$

where $\theta^*(t)$ is the exact solution of (3.339).

Now we are in a position to prove the following result.

Theorem 3.47. *Let conditions* (A1)–(A3) *be fulfilled; then* (3.336) *(or, equivalently,* (3.337), (3.339)*) possesses the unique solution* $\theta^*(t) \in L_q(0, T; Y)$. *This solution is the limit of the sequence* $\{\theta_j(t)\}$ *from* (3.344) *with the factorial convergence, i.e., with the estimate*

$$\| \delta_j(t) \|_{L_q(0,T;Y)} = \| \theta_j(t) - \theta^*(t) \|_{L_q(0,T;Y)}$$

$$\leq \left(\frac{c^{2jq} T^n}{j!} \right)^{1/q} \| \theta_0(t) - \theta^*(t) \|_{L_q(0,T;Y)}. \tag{3.345}$$

For the solution $\theta^*(t) = \partial_0(t)u(t)$, *the following stability estimate holds:*

$$\int_0^t \| \partial_0(\tau)u(\tau) \|_Y^q d\tau \leq 2^{q/p} \int_0^t \| F(\tau) \|_Y^q e^{2^{q/p} c^{2q}(t-\tau)} d\tau. \tag{3.346}$$

Proof. Using the Hölder inequality and assumptions (A1)–(A3) we have

$$\left(\int_0^t \| V^j(\tau_1)s(\cdot) \|^q d\tau_1 \right)^{1/q}$$

$$= \left(\int_0^t \left\| \partial_0(\tau_1) \int_0^{\tau_1} K(\tau_1 - \tau_2) V^{j-1}(\tau_2)s(\cdot)d\tau_2 \right\|^q d\tau_1 \right)^{1/q}$$

$$\leq c \left(\int_0^t \left(\int_0^{\tau_1} \|K(\tau_1 - \tau_2)\| \cdot \|V^{j-1}(\tau_2)s(\cdot)\| d\tau_2 \right)^q d\tau_1 \right)^{1/q}$$

$$\leq c \left(\int_0^t \left(\int_0^{\tau_1} \|K(\tau_1 - \tau_2)\|^p d\tau_2 \right)^{q/p} \cdot \int_0^{\tau_1} \|V^{j-1}(\tau_2)s(\cdot)\|^q d\tau_2 d\tau_1 \right)^{1/q}$$

$$\leq c^2 \left(\int_0^t \int_0^{\tau_1} \|V^{j-1}(\tau_2)s(\cdot)\|^q d\tau_2 d\tau_1 \right)^{1/q} \leq \cdots$$

$$\leq c^{2j} \left(\int_0^t \int_0^{\tau_1} \cdots \int_0^{\tau_j} \|s(\tau_{j+1})\|^q d\tau_{j+1} \ldots d\tau_2 d\tau_1 \right)^{1/q}$$

$$\leq c^{2j} \left(\int_0^t \int_0^{\tau_1} \cdots \int_0^{\tau_{j-1}} d\tau_j \ldots d\tau_2 d\tau_1 \right)^{1/q} \left(\int_0^t \|s(\tau_{j+1})\|^q d\tau_{j+1} \right)^{1/q}$$

$$\leq c^{2j} \left(\frac{t^j}{j!} \right)^{1/q} \left(\int_0^t \|s(\tau_{j+1})\|^q d\tau_{j+1} \right)^{1/q} \qquad \forall s(t) \in L_q(0,T;Y),$$

which means that the spectral radius of the operator $V : L_q(0,T;Y) \to L_q(0,T;Y)$ is equal to zero. The general theory of the Volterra integral equations (see, for example, [36, 34]) yields the existence and uniqueness of the solution of (3.336).

Applying the Hölder inequality to the equation

$$\delta_j(t) = \partial_0(t) \int_0^t K(t - \lambda)\delta_{j-1}(\lambda)d\lambda$$

analogously as above, we deduce that

$$\left(\int_0^t \|\delta_j(\tau_1)\|^q d\tau_1 \right)^{1/q} \leq c^2 \left(\int_0^t \int_0^{\tau_1} \|\delta_{j-1}(\tau_2)\|^q d\tau_2 d\tau_1 \right)^{1/q} \leq \cdots$$

$$\leq c^{2j} \left(\frac{t^j}{j!} \right)^{1/q} \left(\int_0^t \|\delta_0(\tau_{j+1})\|^q d\tau_{j+1} \right)^{1/q},$$

from which the factorial convergence (3.345) follows.

Further, let us prove the stability of the solution of (3.336) with respect to the right-hand side. Using the Hölder inequality for integrals we obtain

$$\left(\int_0^t \|\partial_0(\tau)u(\tau)\|_Y^q d\tau \right)^{1/q}$$

$$\leq \left(\int_0^t \|F(\tau)\|_Y^q d\tau \right)^{1/q} + \left\{ \int_0^t \left\| \partial_0(\tau) \int_0^\tau A^{(1)} e^{-A^{(1)}(\tau-\lambda)} B \partial_0(\lambda) u(\lambda) d\lambda \right\|_Y^q d\tau \right\}^{1/q}$$

$$\leq \left(\int_0^t \|F(\tau)\|_Y^q d\tau \right)^{1/q}$$

$$+ \left\{ \int_0^t \|\partial_0(\tau)\|^q \left(\int_0^\tau \|A^{(1)} e^{-A^{(1)}(\tau-\lambda)} B\|_{Y \to X} \|\partial_0(\lambda)u(\lambda)\|_Y d\lambda \right)^q d\tau \right\}^{1/q}.$$

Applying condition (3.340) to the last summand and then again the Hölder inequality, we arrive at the estimate

$$\left(\int_0^t \|\partial_0(\tau)u(\tau)\|_Y^q d\tau \right)^{1/q} \le \left(\int_0^t \|F(\tau)\|_Y^q d\tau \right)^{1/q}$$

$$+ c \left\{ \int_0^t \left(\int_0^\tau \|A^{(1)}e^{-A^{(1)}(\tau-\lambda)}B\|_{Y\to X}^p d\lambda \right)^{q/p} \int_0^\tau \|\partial_0(\lambda)u(\lambda)\|_Y^q d\lambda d\tau \right\}^{1/q}.$$

Due to (3.341), we further obtain

$$\left(\int_0^t \|\partial_0(\tau)u(\tau)\|_Y^q d\tau \right)^{1/q}$$

$$\le \left(\int_0^t \|F(\tau)\|_Y^q d\tau \right)^{1/q} + c^2 \left\{ \int_0^t \int_0^\tau \|\partial_0(\lambda)u(\lambda)\|_Y^q d\lambda d\tau \right\}^{1/q}.$$

The well-known inequality $(a+b)^q \le 2^{q/p}(a^q+b^q)$ and the last estimate imply

$$\int_0^t \|\partial_0(\tau)u(\tau)\|_Y^q d\tau \le 2^{q/p} \left[\int_0^t \|F(\tau)\|_Y^q d\tau + c^{2q} \int_0^t \int_0^\tau \|\partial_0(\lambda)u(\lambda)\|_Y^q d\lambda d\tau \right].$$

Now the Gronwall lemma [10, 7] yields (3.346).

The proof is complete.   □

Given the solution $\theta^*(t) = \partial_0(t)u(t)$ of integral equation (3.336), problem (3.317) takes the form

$$\frac{du(t)}{dt} + Au(t) = f(t),$$

$$\partial_1 u(t) = g_1(t), \quad u(0) = u_0, \tag{3.347}$$

with a known function $g_1(t) = \theta^*(t) + g(t)$, and its solution is given by (see (3.335))

$$u(t) = v(t) + \int_0^t \frac{\partial}{\partial t} \left\{ e^{-A^{(1)}(t-\lambda)}Bg_1(\lambda) \right\} d\lambda. \tag{3.348}$$

The solution can be computed with exponential accuracy by algorithms from subsection 3.1.4 (see also [27, 18]).

3.4.4   Generalization to a parameter-dependent operator. Existence and uniqueness result

Let $(X, \|\cdot\|)$, $(\mathcal{W}, \|\cdot\|_{\mathcal{W}})$, $(Y, \|\cdot\|_Y)$ be three Banachspaces and $\mathcal{W} \subset X$. For each $t \in [0,T]$ we have a densely defined closed linear unbounded operator $A(t) \colon \mathcal{W} \to X$

and linear bounded operator $\partial_0(t) : W \to Y$. We suppose that the domain $D(A(t)) = D(A) \subset W$ is independent of t.

Let us consider the problem

$$
\begin{aligned}
\frac{du(t)}{dt} + A(t)u(t) &= f(t), \\
\partial_1 u(t) + \partial_0(t)u(t) &= g(t), \quad t \in (0, T], \\
u(0) &= u_0,
\end{aligned}
\tag{3.349}
$$

where u_0 is a given vector, $f(t)$ is a given vector-valued function, and the operator $\partial_0(t)$ is the product of two operators

$$
\partial_0(t) = \mu(t)\partial_0,
\tag{3.350}
$$

with $\partial_0 : D(A) \to Y$, $\mu(t) : Y \to Y$. We suppose that problem (3.349) possesses a unique solution $u(t)$ for all $t \in (0, T)$ for input data f, g, u_0 from a set including the elements $f = 0$, $g = 0$, $u_0 = 0$.

We choose a mesh $\omega_n = \{t_k, \ k = 1, \ldots, n\}$ of n various points on $[0, T]$ and set

$$
\begin{aligned}
\overline{A}(t) &= A_k = A(t_k), \quad t \in (t_{k-1}, t_k], \\
\overline{\mu}(t) &= \mu_k = \mu(t_k), \quad t \in (t_{k-1}, t_k].
\end{aligned}
\tag{3.351}
$$

On each subinterval $(t_{k-1}, t_k]$, we define the operator $A_k^{(2)}$ with a t-independent domain by

$$
\begin{aligned}
D(A_k^{(2)}) &= \{u \in D(A) : \ \partial_1 u + \mu_k \partial_0 u = 0\}, \\
A_k^{(2)} u &= A_k u \quad \forall u \in D(A_k^{(2)})
\end{aligned}
\tag{3.352}
$$

and the operator $B_k^{(1)} : Y \to D(A)$ by

$$
\begin{aligned}
A_k(B_k^{(1)} y) &= 0, \\
(\partial_1 + \mu_k \partial_0) B_k^{(1)} y &= y.
\end{aligned}
\tag{3.353}
$$

For all $t \in [0, T)$, we define the operators

$$
\begin{aligned}
A^{(2)}(t) &= A_k^{(2)}, \quad t \in (t_{k-1}, t_k], \\
B^{(1)}(t) &= B_k^{(1)}, \quad t \in (t_{k-1}, t_k], \quad \forall k = 1, \ldots, n
\end{aligned}
\tag{3.354}
$$

(existence of $B_k^{(1)}$ can be shown analogously to existence of B from subsection 3.4.2).

Further, we accept the following hypotheses (see subsection 3.4.6 for examples):

(B1) We suppose the operator $A^{(2)}(t)$ to be strongly positive (see (3.323)).

This assumption implies that there exist a positive constant c and a fixed κ such that (see, e.g., [14, p. 103])

$$\|[A^{(2)}(t)]^\kappa e^{-sA^{(2)}(t)}\| \le cs^{-\kappa}, \quad s > 0, \quad \kappa > 0.$$

(B2) There exists a real positive ω such that

$$\|e^{-sA^{(2)}(t)}\| \le e^{-\omega s} \quad \forall s, t \in [0, T]$$

(see [54, Corollary 3.8, p. 12] for the corresponding assumptions on $A(t)$).

We also assume that the following conditions hold:

(B3) $\|[A^{(2)}(t) - A^{(2)}(s)][A^{(2)}(t)]^{-\gamma}\| \le c|t - s| \quad \forall t, s, \ 0 \le \gamma \le 1;$

(B4) $\|[A^{(2)}(t)]^\beta [A^{(2)}(s)]^{-\beta} - I\| \le c|t - s| \quad \forall t, s \in [0, T], \quad \beta \in (0, 1).$

(B5) The operator $\mu(t)\partial_0$ satisfies the following conditions:

$$\|\mu(t) - \mu(t')\|_{Y \to Y} \le M|t - t'|,$$
$$\|\partial_0\|_{X \to Y} \le c.$$

(B6) For some $p \ge 1$ and $\gamma \ge 0$, there holds

$$\left[\int_0^t \|[A^{(2)}(\eta)]^{1+\gamma} e^{-A^{(2)}(\eta)(t-\lambda)} B^{(1)}(\eta)\|_{Y \to X}^p d\lambda \right]^{1/p} \le c \quad \forall t, \ \eta \in [0, T].$$

Let us rewrite problem (3.349) in the equivalent form (so-called prediscretization)

$$\frac{du}{dt} + \overline{A}(t)u = [\overline{A}(t) - A(t)]u(t) + f(t),$$
$$[\partial_1 + \overline{\mu}(t)\partial_0]u(t) = -[\mu(t) - \overline{\mu}(t)]\partial_0 u(t) + g(t), \quad t \in [0, T], \qquad (3.355)$$
$$u(0) = u_0.$$

Note that now all operators on the left-hand side of these equations are constant on each subinterval and piecewise constant on the whole interval $[0, T)$.

From (3.355) analogously to (3.335), (3.337) we deduce

$$u(t) = e^{-A_k^{(2)}(t-t_{k-1})} u(t_{k-1})$$
$$+ \int_{t_{k-1}}^t e^{-A_k^{(2)}(t-\tau)} \{[A_k - A(\tau)]u(\tau) + f(\tau)\} d\tau$$

$$+ \int_{t_{k-1}}^{t} A_k^{(2)} e^{-A_k^{(2)}(t-\lambda)} B_k^{(1)} \{ -[\mu(\lambda) - \mu_k] \partial_0 u(\lambda) + g(\lambda) \} d\lambda, \quad (3.356)$$

$$\partial_0 u(t) = \partial_0 e^{-A_k^{(2)}(t-t_{k-1})} u(t_{k-1})$$

$$+ \partial_0 \int_{t_{k-1}}^{t} e^{-A_k^{(2)}(t-\tau)} \{ [A_k - A(\tau)] u(\tau) + f(\tau) \} d\tau$$

$$+ \partial_0 \int_{t_{k-1}}^{t} A_k^{(2)} e^{-A_k^{(2)}(t-\lambda)} B_k^{(1)} \{ -[\mu(\lambda) - \mu_k] \partial_0 u(\lambda) + g(\lambda) \} d\lambda,$$

$$t \in [t_{k-1}, t_k], \quad k = 1, \dots, n.$$

Thus, with the Duhamel-like technique, we have obtained the system of two integral equations with respect to the unknown functions $u(t)$ and $\partial_0 u(t)$ which is equivalent to (3.349). This system is the starting point for our forthcoming investigations and for a numerical algorithm.

To prove the existence and uniqueness result it is sufficient to choose in the framework above $n = 1$, $t_0 = 0$, $t_1 = T$, $\overline{A}(t) = A(0) = A$, and $\overline{\mu}(t) = \mu(T)$. In addition, we introduce the vectors

$$\mathcal{U}(t) = \begin{pmatrix} u(t) \\ \theta(t) \end{pmatrix}, \quad \mathcal{U}_n(t) = \begin{pmatrix} u_n(t) \\ \theta_n(t) \end{pmatrix}, \quad \mathcal{F} = \begin{pmatrix} F_1(t) \\ F_2(t) \end{pmatrix} \quad (3.357)$$

and the matrices

$$\mathcal{K}(t, \tau) = \begin{pmatrix} K_{11}(t, \tau) & K_{12}(t, \tau) \\ K_{21}(t, \tau) & K_{22}(t, \tau) \end{pmatrix},$$

$$\mathcal{D} = \begin{pmatrix} I & 0 \\ 0 & \partial_0 \end{pmatrix}, \quad \mathcal{E}_\gamma = \begin{pmatrix} [A^{(2)}]^\gamma & 0 \\ 0 & I \end{pmatrix}, \quad (3.358)$$

with

$$K_{11}(t, \tau) = K_{21}(t, \tau) = -e^{-A^{(2)}(t-\tau)}[A(\tau) - A],$$

$$K_{12}(t, \tau) = K_{22}(t, \tau) = -A^{(2)} e^{-A^{(2)}(t-\tau)} B^{(1)}(t)[\mu_k - \mu(\tau)],$$

$$F_1(t) = e^{-A^{(2)}t} u_0 + \int_0^t e^{-A^{(2)}(t-\tau)} f(\tau) d\tau \quad (3.359)$$

$$+ \int_0^t A^{(2)} e^{-A^{(2)}(t-\lambda)} B^{(1)} g(\lambda) d\lambda, \quad F_2(t) = \partial_0 F_1(t).$$

We also introduce the space $\mathcal{Y}$ of vectors $\mathcal{U} = (u, v)^T$ with the norm

$$\|\mathcal{U}\|_\mathcal{Y} = \max\{\|u\|_{L_q(0,T;D([A^{(2)}]^\gamma))}, \|v\|_{L_q(0,T;Y)}\}, \quad (3.360)$$

and we equip the space of matrices $\mathcal{K}(t, \tau)$ from (3.358) with the matrix norm

$$\|\mathcal{K}\|_\infty = \max\{\|K_{1,1}\|_{X \to X} + \|K_{1,2}\|_{Y \to X}, \|K_{2,1}\|_{X \to X} + \|K_{2,2}\|_{Y \to X}\}, \quad (3.361)$$

which is consistent with the vector norm (3.360). Now the system (3.356) with $k = 1$, $t_0 = 0, t_1 = T$ can be written in the form

$$\mathcal{U}(t) = \mathcal{D} \int_0^t \mathcal{K}(t, \tau)\mathcal{U}(\tau)d\tau + \mathcal{F}(t). \tag{3.362}$$

The fixed point iteration for the system (3.362) is given by

$$\mathcal{U}_{j+1}(t) = \mathcal{D} \int_0^t \mathcal{K}(t, \tau)\mathcal{U}_j(\tau)d\tau + \mathcal{F}(t), \quad j = 0, 1, \ldots, \quad \mathcal{U}_0(t) = 0. \tag{3.363}$$

Since the operators $\mathcal{E}_\gamma$ and $\mathcal{D}(t)$ commute, we get from (3.363) that

$$\mathcal{U}_{\gamma,j+1}(t) = \mathcal{D} \int_0^t \mathcal{E}_\gamma \mathcal{K}(t, \tau)\mathcal{E}_\gamma^{-1}\mathcal{U}_{\gamma,j}(\tau)d\tau + \mathcal{F}_\gamma(t), \quad j = 0, 1, \ldots,$$
$$\mathcal{U}_0(t) = 0, \tag{3.364}$$

where $\mathcal{U}_{\gamma,j}(t) = \mathcal{E}_\gamma \mathcal{U}_j(t)$ and $\mathcal{F}_\gamma(t) = \mathcal{E}_\gamma \mathcal{F}(t)$.

Now we are in a position to formulate the following result (the proof is completely analogous to the one of Theorem 3.47).

Theorem 3.48. *Let us assume that the conditions* (B1)–(B6) *are fulfilled; then the system of equations* (3.356) *possesses the unique solution* $\mathcal{U}_\gamma^*(t)$ *in* $\mathcal{Y}$. *This solution is the limit of the sequence* $\mathcal{U}_{\gamma,j}(t)$ *from* (3.364) *with the factorial convergence characterized by the estimate*

$$\|\mathcal{U}_{\gamma,j}(t) - \mathcal{U}_\gamma^*(t)\|_{\mathcal{Y}} \leq \left(\frac{c^{2jq}T^j}{j!} \right)^{1/q} \|\mathcal{U}_{\gamma,0}(t) - \mathcal{U}_\gamma^*(t)\|_{\mathcal{Y}}. \tag{3.365}$$

For the solution $\mathcal{U}_\gamma^*(t)$ *the following stability estimate holds:*

$$\int_0^t \|\mathcal{U}_\gamma^*(\tau)\|_{\mathcal{Y}}^q d\tau \leq 2^{q/p} \int_0^t \|\mathcal{F}(\tau)\|_{\mathcal{Y}}^q e^{2^{q/p} c^{2q}(t-\tau)} d\tau. \tag{3.366}$$

Proof. Let us introduce the linear operator $\mathcal{V}(t)$ by

$$\mathcal{V}(t)\mathcal{U}(\cdot) = \mathcal{D}(t) \int_0^t \mathcal{E}_\gamma \mathcal{K}(t, \tau)\mathcal{E}_\gamma^{-1}\mathcal{U}(\tau)d\tau.$$

Using assumptions (B1) and (B3), we get the following for the element $[\mathcal{E}_\gamma \mathcal{K}\mathcal{E}_\gamma^{-1}]_{1,1}$ of the matrix $\mathcal{E}_\gamma \mathcal{K}\mathcal{E}_\gamma^{-1}$ with the indexes $(1, 1)$:

$$\left\| [\mathcal{E}_\gamma \mathcal{K}(\tau_1, \tau_2)\mathcal{E}_\gamma^{-1}]_{1,1} \right\| = \left\| \left[A^{(2)} \right]^\gamma e^{-A^{(2)}(\tau_1 - \tau_2)} [A(\tau_2) - A] \left[A^{(2)} \right]^{-\gamma} \right\|$$

$$\leq \left\| \left[A^{(2)} \right]^\gamma e^{-A^{(2)}(\tau_1 - \tau_2)} \right\| \left\| [A(\tau_2) - A] \left[A^{(2)} \right]^{-\gamma} \right\| \leq c^2(\tau_1 - \tau_2)^{-\gamma}(T - \tau_2).$$

This estimate implies

$$\int_0^{\tau_1} \| \left[\mathcal{E}_\gamma K(\tau_1, \tau_2) \mathcal{E}_\gamma^{-1} \right]_{1,1} \|^p d\tau_2 \le c^{2p} \int_0^{\tau_1} (\tau_1 - \tau_2)^{-\gamma p} (T - \tau_2)^p d\tau_2$$

$$\le c^{2p} \tau_1^{1-\gamma p} T^p \int_0^1 (1 - \eta)^{-\gamma p} d\eta = c^{2p} \tau_1^{1-\gamma p} \frac{T^p}{1 - \gamma p} \le C^p,$$

with some new constant C. This expression remains bounded for $\gamma p \in [0, 1)$.

Assumption (B6) implies

$$\left(\int_0^{\tau_1} \| \left[\mathcal{E}_\gamma K(\tau_1, \tau_2) \mathcal{E}_\gamma^{-1} \right]_{1,2} \|^p d\tau_2 \right)^{1/p}$$

$$= \left(\int_0^{\tau_1} \left\| \left[A^{(2)} \right]^{\gamma+1} e^{-A^{(2)}(\tau_1 - \tau_2)} B^{(1)} \right\|^p d\tau_2 \right)^{1/p} \le C,$$

$$\left(\int_0^{\tau_1} \| \left[\mathcal{E}_\gamma K(\tau_1, \tau_2) \mathcal{E}_\gamma^{-1} \right]_{2,2} \|^p d\tau_2 \right)^{1/p}$$

$$= \left(\int_0^{\tau_1} \| A^{(2)} e^{-A^{(2)}(\tau_1 - \tau_2)} B^{(1)} \|^p d\tau_2 \right)^{1/p} \le C.$$

Using (B2) and (B3) we have

$$\left(\int_0^{\tau_1} \| \left[\mathcal{E}_\gamma K(\tau_1, \tau_2) \mathcal{E}_\gamma^{-1} \right]_{2,1} \|^p d\tau_2 \right)^{1/p}$$

$$= \left(\int_0^{\tau_1} \left\| e^{-A^{(2)}(\tau_1 - \tau_2)} [A(\tau_2) - A] \left[A^{(2)} \right]^{-\gamma} \right\|^p d\tau_2 \right)^{1/p}$$

$$\le \left(\int_0^{\tau_1} \left\| [A(\tau_2) - A] \left[A^{(2)} \right]^{-\gamma} \right\|^p d\tau_2 \right)^{1/p} \le c \left(\int_0^{\tau_1} (T - \tau_2)^p d\tau_2 \right)^{1/p} \le C.$$

Therefore

$$\| \mathcal{E}_\gamma K(\tau_1, \tau_2) \mathcal{E}_\gamma^{-1} \|_\infty \le 2 \max_{i,j} \| \left[\mathcal{E}_\gamma K(\tau_1, \tau_2) \mathcal{E}_\gamma^{-1} \right]_{i,j} \| \le C.$$

Assumption (B5) yields

$$\| \mathcal{D}(\tau_1) \|_1 \le \max \{1, \| \partial_0(\tau) \| \} \le C.$$

Now, using estimates (3.4.4) and (3.4.4) we get, for the n-th power of the operator $\mathcal{V}(t)$,

$$\left(\int_0^t \| \mathcal{V}^j(\tau_1) s(\cdot) \|^q d\tau_1 \right)^{1/q}$$

$$= \left(\int_0^t \left\| \mathcal{D}(\tau_1) \int_0^{\tau_1} \mathcal{E}_\gamma K(\tau_1, \tau_2) \mathcal{E}_\gamma^{-1} \mathcal{V}^{j-1}(\tau_2) s(\cdot) d\tau_2 \right\|^q d\tau_1 \right)^{1/q}$$

$$\leq \left(\int_0^t \|\mathcal{D}(\tau_1)\|_\infty^q \left(\int_0^{\tau_1} \|\mathcal{E}_\gamma K(\tau_1,\tau_2)\mathcal{E}_\gamma^{-1}\mathcal{V}^{j-1}(\tau_2)s(\cdot)\|_\infty d\tau_2 \right)^q d\tau_1 \right)^{1/q}$$

$$\leq C \left(\int_0^t \left(\int_0^{\tau_1} \|\mathcal{E}_\gamma K(\tau_1,\tau_2)\mathcal{E}_\gamma^{-1}\|_\infty \|\mathcal{V}^{j-1}(\tau_2)s(\cdot)\|_\infty d\tau_2 \right)^q d\tau_1 \right)^{1/q}$$

$$\leq C \left(\int_0^t \left(\int_0^{\tau_1} \|\mathcal{E}_\gamma K(\tau_1,\tau_2)\mathcal{E}_\gamma^{-1}\|_\infty^p d\tau_2 \right)^{q/p} \cdot \int_0^{\tau_1} \|\mathcal{V}^{j-1}(\tau_2)s(\cdot)\|_\infty^q d\tau_2 d\tau_1 \right)^{1/q}$$

$$\leq C^2 \left(\int_0^t \int_0^{\tau_1} \|\mathcal{V}^{j-1}(\tau_2)s(\cdot)\|_\infty^q d\tau_2 d\tau_1 \right)^{1/q} \leq \cdots$$

$$\leq C^{2j} \left(\int_0^t \int_0^{\tau_1} \cdots \int_0^{\tau_j} \|s(\tau_{j+1})\|_\infty^q d\tau_{j+1} \ldots d\tau_2 d\tau_1 \right)^{1/q}$$

$$\leq C^{2j} \left(\int_0^t \int_0^{\tau_1} \cdots \int_0^{\tau_{j-1}} d\tau_j \ldots d\tau_2 d\tau_1 \right)^{1/q} \left(\int_0^t \|s(\tau_{j+1})\|_\infty^q d\tau_{j+1} \right)^{1/q}$$

$$\leq C^{2j} \left(\frac{t^j}{j!} \right)^{1/q} \left(\int_0^t \|s(\tau_{j+1})\|_\infty^q d\tau_{j+1} \right)^{1/q} \quad \forall s(t) \in \mathcal{Y}.$$

For the difference

$$\Delta_j(t) = \mathcal{U}_j(t) - \mathcal{U}^*(t)$$

between the j-th iteration and the exact solution, we have the equation

$$\Delta_j(t) = \mathcal{D}(t) \int_0^t \mathcal{E}_\gamma K(t,\tau)\mathcal{E}_\gamma^{-1}\Delta_{j-1}(\tau)d\tau.$$

Applying the Hölder inequality analogously as above, we get

$$\left(\int_0^t \|\Delta_j(\tau_1)\|_\infty^q d\tau_1 \right)^{1/q} \leq c^2 \left(\int_0^t \int_0^{\tau_1} \|\Delta_{j-1}(\tau_2)\|_\infty^q d\tau_2 d\tau_1 \right)^{1/q} \leq \cdots$$

$$\leq c^{2j} \left(\frac{t^j}{j!} \right)^{1/q} \left(\int_0^t \|\Delta_0(\tau_{j+1})\|_\infty^q d\tau_{j+1} \right)^{1/q},$$

from which the factorial convergence (3.365) follows.

Finally, let us prove the stability of the solution of (3.361) with respect to the right-hand side. Using the Hölder inequality for integrals, we obtain

$$\left(\int_0^t \|\mathcal{U}_\gamma(\tau)\|_\infty^q d\tau \right)^{1/q} \leq \left(\int_0^t \|\mathcal{F}(\tau)\|_\infty^q d\tau \right)^{1/q}$$

$$+ \left\{ \int_0^t \left\| \mathcal{D}(\tau_1) \int_0^{\tau_1} \mathcal{E}_\gamma K(\tau_1,\tau_2)\mathcal{E}_\gamma^{-1}\mathcal{U}_\gamma(\tau_2)d\tau_2 \right\|_\infty^q d\tau_1 \right\}^{1/q}.$$

Using estimates as above we obtain

$$\left(\int_0^t \|\mathcal{U}_\gamma(\tau)\|_\infty^q d\tau \right)^{1/q} \le c^2 \left(\int_0^t \int_0^{\tau_1} \|\mathcal{U}_\gamma(\tau_2)\|_\infty^q d\tau_2 d\tau_1 \right)^{1/q} + \left(\int_0^t \|\mathcal{F}(\tau)\|_\infty^q d\tau \right)^{1/q}.$$

This estimate together with the inequality $(a+b)^q \le 2^{q/p}(a^q + b^q)$ implies

$$\int_0^t \|\mathcal{U}_\gamma(\tau)\|_\infty^q d\tau \le 2^{q/p} \left[c^{2q} \int_0^t \int_0^{\tau_1} \|\mathcal{U}_\gamma(\tau_2)\|_\infty^q d\tau_2 d\tau_1 + \int_0^t \|\mathcal{F}(\tau)\|_\infty^q d\tau \right].$$

Now, the Gronwall lemma yields (3.366).

The proof is complete. $\qquad\square$

3.4.5 Numerical algorithm

To construct a discrete approximation of (3.356) we use the Chebyshev interpolation on the interval $[-1, 1]$ (if it is not the case, one can reduce the problem (3.349) to this interval by the variable transform $t = 2t'/T - 1$, $t \in [-1, 1]$, $t' \in [0, T]$). We choose a mesh $w_n = \{t_k = -\cos\frac{(2k-1)\pi}{2n},\ k = 1, \dots, n\}$ on $[-1, 1]$ of n zeros of Chebyshev orthogonal polynomial $T_n(t)$ and set $\tau_k = t_k - t_{k-1}$. It is well known that (see [63, Chapter 6, Theorem 6.11.12], [64, p. 123])

$$t_{\nu+1} - t_\nu < \frac{\pi}{n}, \quad \nu = 1, \dots, n,$$

$$\tau = \tau_{max} = \max_{1 \le k \le n} \tau_k < \frac{\pi}{n}. \tag{3.367}$$

Let

$$P_{n-1}(t; u) = P_{n-1}u = \sum_{j=1}^n u(t_j) L_{j,n-1}(t),$$

$$\tag{3.368}$$

$$P_{n-1}(t; \partial_0 u) = P_{n-1}(\partial_0 u) = \sum_{j=1}^n \partial_0 u(t_j) L_{j,n-1}(t)$$

be the Lagrange interpolation polynomials for $u(t)$ and $\partial_0 u(t)$ on the mesh w_n, where $L_{j,n-1} = \frac{T_n(t)}{T_n'(t_j)(t-t_j)}$, $j = 1, \dots, n$, are the Lagrange fundamental polynomials. For a given vector $v = (v_1, \dots, v_n)$, we introduce the interpolation polynomial

$$P_{n-1}(t; v) = P_{n-1}y = \sum_{j=1}^n v_j L_{j,n-1}(t), \tag{3.369}$$

so that $P_{n-1}(t_j; v) = v_j$, $j = 1, 2, \dots, n$. Let $x = (x_1, x_2, \dots, x_n)$ and $y = (y_1, y_2, \dots, y_n)$ be the approximating vectors for $U = (u(t_1), u(t_2), \dots, u(t_n))$ and $\partial U = (\partial_0 u(t_1), \partial_0 u(t_2), \dots, \partial_0 u(t_n))$, respectively; i.e., x_k approximates $u(t_k)$, and y_k approximates $\partial_0 u(t_k)$. Substituting $P_{n-1}(\eta; x)$ for $u(\eta)$ and $P_{n-1}(\eta; y)$ for $\partial_0 u(\lambda)$ and then setting $t = t_k$ in (3.356), we arrive at the following system of linear equations with respect to the unknown elements x_k, y_k:

$$
\begin{aligned}
x_k &= e^{-A_k^{(2)} \tau_k} x_{k-1} + \sum_{j=1}^{n} \alpha_{kj} x_j + \sum_{j=1}^{n} \beta_{kj} y_j + \phi_k, \\
y_k &= \partial_0 e^{-A_k^{(2)} \tau_k} x_{k-1} + \partial_0 \sum_{j=1}^{n} \alpha_{kj} x_j + \partial_0 \sum_{j=1}^{n} \beta_{kj} y_j + \partial_0 \phi_k, \\
&k = 1, \ldots, n, \quad x_0 = u_0, \quad y_0 = \partial_0 u_0,
\end{aligned}
\tag{3.370}
$$

which represents our algorithm. Here we use the notation

$$
\begin{aligned}
\alpha_{kj} &= \int_{t_{k-1}}^{t_k} e^{-A_k^{(2)}(t_k - \eta)} [A_k - A(\eta)] L_{j,n-1}(\eta) d\eta, \\
\beta_{kj} &= \int_{t_{k-1}}^{t_k} A_k^{(2)} e^{-A_k^{(2)}(t_k - \lambda)} B_k^{(1)} [\mu(\lambda) - \mu_k] L_{j,n-1}(\lambda) d\lambda, \\
\phi_k &= \int_{t_{k-1}}^{t_k} e^{-A_k^{(2)}(t_k - \eta)} f(\eta) d\eta + \int_{t_{k-1}}^{t_k} A_k^{(2)} e^{-A_k^{(2)}(t_k - \lambda)} B_k^{(1)} g(\lambda) d\lambda.
\end{aligned}
\tag{3.371}
$$

Furthermore, for the sake of simplicity we analyze this algorithm for the particular case of problem (3.349), where operator $A(t)$ is independent of t, i.e., $A(t) \equiv A$. In this case we have $\alpha_{kj} = 0$, and system (3.370) takes the form

$$
\begin{aligned}
x_k &= e^{-A_k^{(2)} \tau_k} x_{k-1} + \sum_{j=1}^{n} \beta_{kj} y_j + \phi_k, \\
y_k &= \partial_0 e^{-A_k^{(2)} \tau_k} x_{k-1} + \partial_0 \sum_{j=1}^{n} \beta_{kj} y_j + \partial_0 \phi_k, \\
&k = 1, \ldots, n, \quad x_0 = u_0, \quad y_0 = \partial_0 u_0.
\end{aligned}
\tag{3.372}
$$

Remark 3.49. Under the assumption that $f(t), \mu(t), g(t)$ are polynomials, the calculation of the operators β_{kj} and the elements ϕ_k can be reduced to the calculation of integrals of the kind $I_s = \int_{t_{k-1}}^{t_k} e^{-A_k^{(2)}(t_k - \lambda)} \lambda^s d\lambda$, which can be found by a simple recurrence algorithm: $I_l = -l[A_k^{(2)}]^{-1} I_{l-1} + [A_k^{(2)}]^{-1}(t_k^l I - t_{k-1}^l e^{-A_k^{(2)} \tau_k})$, $l = 1, 2, \ldots, s$, $I_0 = [A_k^{(2)}]^{-1}(I - e^{-A_k^{(2)} \tau_k})$, where the operator exponentials can be computed by the exponentially convergent algorithm from the subsection 3.1.4.

For the errors $z_x = (z_{x,1}, \ldots, z_{x,n})$, $z_y = (z_{y,1}, \ldots, z_{y,n})$, with $z_{x,k} = u(t_k) - x_k$ and $z_{y,k} = \partial_0 u(t_k) - y_k$, we have the relations

$$z_{x,k} = e^{-A_k \tau_k} z_{x,k-1} + \sum_{j=1}^{n} \beta_{kj} z_{y,j} + \psi_{y,k},$$

$$z_{y,k} = \partial_0 e^{-A_k \tau_k} z_{y,k-1} + \partial_0 \sum_{j=1}^{n} \beta_{kj} z_{y,j} + \partial_0 \psi_{y,k}, \qquad (3.373)$$

$$k = 1, \ldots, n, \quad z_{x,0} = 0, \quad z_{y,0} = 0,$$

where

$$\psi_{y,k} = \int_{t_{k-1}}^{t_k} A_k^{(2)} e^{-A_k^{(2)}(t_k - \eta)} B_k^{(1)} [\mu_k - \mu(\eta)][\partial_0 u(\eta) - P_n(\eta; \partial_0 u)] d\eta. \qquad (3.374)$$

To represent algorithm (3.370) in a block-matrix form, we introduce a matrix like the one in (3.165),

$$S = \{s_{i,k}\}_{i,k=1}^n = \begin{pmatrix} E_X & 0 & 0 & \cdot & \cdot & 0 & 0 \\ -e^{-A_2^{(2)}\tau_2} & E_X & 0 & \cdot & \cdot & 0 & 0 \\ 0 & -e^{-A_3^{(2)}\tau_3} & E_X & \cdot & \cdot & 0 & 0 \\ \cdot & \cdot & \cdot & \cdot & \cdot & \cdot & \cdot \\ 0 & 0 & 0 & \cdot & \cdot & -e^{-A_n^{(2)}\tau_n} & E_X \end{pmatrix},$$

with E_X being the identity operator in X, the matrix $D = \{\beta_{k,j}\}_{k,j=1}^n$, and the vectors

$$x = \begin{pmatrix} x_1 \\ \cdot \\ \cdot \\ \cdot \\ x_n \end{pmatrix}, \quad f_x = \begin{pmatrix} \phi_1 \\ \cdot \\ \cdot \\ \cdot \\ \phi_n \end{pmatrix}, \quad \tilde{f}_x = \begin{pmatrix} e^{-A_1^{(2)}\tau_1} u_0 \\ 0 \\ \cdot \\ \cdot \\ 0 \end{pmatrix}, \quad \psi_y = \begin{pmatrix} \psi_{y,1} \\ \cdot \\ \cdot \\ \psi_{y,n} \end{pmatrix},$$

$$f_y = \begin{pmatrix} \partial_0 \phi_1 \\ \cdot \\ \cdot \\ \partial_0 \phi_n \end{pmatrix}, \quad \tilde{f}_y = \begin{pmatrix} \partial_0 e^{-A_1^{(2)}\tau_1} u_0 \\ 0 \\ \cdot \\ \cdot \\ 0 \end{pmatrix}, \quad \psi_y^{(0)} = \begin{pmatrix} \partial_0 \psi_{y,1} \\ \cdot \\ \cdot \\ \partial_0 \psi_{y,n} \end{pmatrix}.$$

$$(3.375)$$

It is easy to see that for the (left) inverse

$$S^{-1} = \begin{pmatrix} E_X & 0 & \cdots & 0 & 0 \\ s_1 & E_X & \cdots & 0 & 0 \\ s_2 s_1 & s_2 & \cdots & 0 & 0 \\ \cdot & \cdot & \cdots & \cdot & \cdot \\ s_{n-1} \cdots s_1 & s_{n-1} \cdots s_2 & \cdots & s_{n-1} & E_X \end{pmatrix}, \qquad (3.376)$$

where $s_k = -e^{-A_{k+1}^{(2)}\tau_{k+1}}$, it holds that

$$
S^{-1}S = \begin{pmatrix} E_X & 0 & \cdots & 0 \\ 0 & E_X & \cdots & 0 \\ . & . & \cdots & . \\ 0 & 0 & \cdots & E_X \end{pmatrix}.
\tag{3.377}
$$

Remark 3.50. Using results of [27, 18, 19] one can get a parallel and sparse approximation with an exponential convergence rate of the operator exponentials contained in S^{-1} and, as a consequence, a parallel and sparse approximation of S^{-1}.

The first and second equations of (3.370) can be written in the matrix form as

$$
\begin{aligned}
Sx &= Dy + f_x + \tilde{f}_x, \\
y &= \Lambda[(I_X - S)x + Dy + f_x + \tilde{f}_x],
\end{aligned}
\tag{3.378}
$$

where

$$
\Lambda = \mathrm{diag}\,[\partial_0, \ldots, \partial_0], \quad I_X = \mathrm{diag}(E_X, \ldots, E_X).
\tag{3.379}
$$

The errors z_x, z_y satisfy the equations

$$
\begin{aligned}
Sz_x &= Dz_y + \psi_y, \\
z_y &= \Lambda\,[(I_X - S)z_x + Dz_y + \psi_y].
\end{aligned}
\tag{3.380}
$$

From the second equation in (3.378) and the second equation in (3.380) we obtain

$$
\begin{aligned}
[I_Y - \Lambda D]y &= \Lambda[(I_X - S)x + f_x + \tilde{f}_x], \\
[I_Y - \Lambda D]z_y &= \Lambda[(I_X - S)z_x + \psi_y],
\end{aligned}
\tag{3.381}
$$

where $I_Y = \mathrm{diag}(E_Y, \ldots, E_Y)$ and E_Y is the identity operator in Y.

For a vector $v = (v_1, v_2, \ldots, v_n)^T$ and a block operator matrix $A = \{a_{ij}\}_{i,j=1}^n$, we introduce the vector norm

$$
|||v||| \equiv |||v|||_\infty = \max_{1 \le k \le n} \|v_k\|
\tag{3.382}
$$

and the consistent matrix norm

$$
|||A||| \equiv |||A|||_\infty = \max_{1 \le i \le n} \sum_{j=1}^n \|a_{i,j}\|.
\tag{3.383}
$$

For further analysis we need the following auxiliary result.

Lemma 3.51. *Under assumptions* (B1)–(B6) *the following estimates hold true:*

$$\||S^{-1}\|| \leq n,$$
$$\||D\|| \leq cn^{-(1+1/q)} \ln n, \quad 1/p + 1/q = 1,$$
$$\||\Lambda\|| \leq c,$$

(3.384)

with a positive constant c independent of n.

Proof. The assumption (B2) yields

$$\||S^{-1}\|| \leq 1 + e^{-\omega\tau} + \cdots + e^{-\omega\tau(n-1)} \leq n.$$

(3.385)

For the matrix D, we have from (B6)

$$\||D\|| \leq \max_{1 \leq k \leq n} \sum_{j=1}^{n} \|\beta_{kj}\|$$

$$= \max_{1 \leq k \leq n} \sum_{j=1}^{n} \left\| \int_{t_{k-1}}^{t_k} A_k^{(2)} e^{-A_k^{(2)}(t_k - \tau)} B_k^{(1)} [\mu(t_k) - \mu(\tau)] L_{j,n-1}(\tau) d\tau \right\|$$

$$\leq \max_{1 \leq k \leq n} \int_{t_{k-1}}^{t_k} \|A_k^{(2)} e^{-A_k^{(2)}(t_k - \tau)} B_k^{(1)}\| \|\mu(t_k) - \mu(\tau)\| \sum_{j=1}^{n} |L_{j,n-1}(\tau)| d\tau$$

$$\leq c\Lambda_n (t_k - t_{k-1}) \int_{t_{k-1}}^{t_k} \|A_k^{(2)} e^{-A_k^{(2)}(t_k - \tau)} B_k^{(1)}\| d\tau \leq cn^{-(1+1/q)} \ln n,$$

(3.386)

where $\Lambda_n = \max_{-1 \leq \tau \leq 1} \sum_{j=1}^{n} |L_{j,n-1}(\tau)|$ is the Lebesgue constant related to the Chebyshev interpolation nodes.

The last estimate is a simple consequence of assumption (B5). The lemma is proved. $\square$

Due to (3.384) for n large enough, there exists the inverse $(I_Y - \Lambda D)^{-1}$, which is bounded by a constant c independent of n; i.e.,

$$\||(I_Y - \Lambda D)^{-1}\|| \leq c.$$

(3.387)

Therefore, we obtain from (3.381)

$$y = [I_Y - \Lambda D]^{-1} \Lambda \left[(I_X - S)x + f_x + \tilde{f}_x \right],$$
$$z_y = [I_Y - \Lambda D]^{-1} \Lambda \left[(I_X - S)z_x + \psi_y \right].$$

(3.388)

Substituting these expressions into the first equation in (3.378) and (3.380), respectively, we obtain

$$Gx = Q(f_x + \tilde{f}_x),$$
$$Gz_x = Q\psi_y,$$

(3.389)

where

$$G = S - D[I_Y - \Lambda D]^{-1}\Lambda(I_X - S),$$
$$Q = D[I_Y - \Lambda D]^{-1}\Lambda + I_X. \tag{3.390}$$

The next lemma presents estimates for G^{-1} and Q.

Lemma 3.52. *Under assumptions of Lemma* 3.51 *there exists* G^{-1}, *and there holds*

$$|||G^{-1}||| \leq cn,$$
$$|||Q||| \leq c \tag{3.391}$$

with some constant independent of n.

Proof. We represent $G = S[I_X - G_1]$ with $G_1 = S^{-1}D[I_Y - \Lambda D]^{-1}\Lambda(I_X - S)$ and estimate $|||G_1|||$. We have

$$|||G_1||| \leq |||S^{-1}||| \cdot |||D||| \cdot |||(I_Y - \Lambda D)^{-1}||| \cdot |||\Lambda||| \cdot |||I_X - S|||,$$

and now Lemma 3.51 implies that

$$|||G_1||| \leq n \, n^{-(1+1/q)}c\ln n = c\frac{\ln n}{n^{1/q}}. \tag{3.392}$$

This estimate guarantees the existence of the bounded inverse operator $(I_X - G_1)^{-1}$, which together with the estimate $|||S^{-1}||| \leq n$ proves the first assertion of the lemma. The second assertion is evident. The proof is complete. □

This lemma and (3.389) imply the following stability estimates:

$$|||x||| \leq cn|||f_x + \tilde{f}_x|||,$$
$$|||z_x||| \leq cn|||\tilde{\psi}_y|||. \tag{3.393}$$

Substituting estimates (3.393) into (3.388) and taking into account the estimates $|||I_X - S||| \leq c$, $|||\Lambda||| \leq c$ as well as (3.387), we obtain

$$|||y||| \leq cn|||f_x + \tilde{f}_x|||,$$
$$|||z_y||| \leq cn|||\psi_y|||. \tag{3.394}$$

Remark 3.53. We have reduced the interval length to $T = 2$, but in general the constants c involved depend on the interval length T.

Let Π_{n-1} be the set of all polynomials in t with vector coefficients of degree less than or equal to $n - 1$. In complete analogy with [5, 63, 64], the following Lebesgue inequality for vector-valued functions can be proved:

$$\|u(\eta) - P_{n-1}(\eta; u)\|_{C[-1,1]} \equiv \max_{\eta \in [-1,1]} \|u(\eta) - P_{n-1}(\eta; u)\| \leq (1 + \Lambda_n)E_n(u),$$

with the error of the best approximation of u by polynomials of degree not greater than $n - 1$,

$$E_n(u) = \inf_{p \in \Pi_{n-1}} \max_{\eta \in [-1,1]} \|u(\eta) - p(\eta)\|.$$

Now we can go to the main result of this section.

Theorem 3.54. *Let the assumptions of Lemma 3.51 hold; then there exists a positive constant c such that:*

1. *for n large enough it holds that*

$$\begin{aligned} \||z_x\|| &\leq cn^{-1/q} \cdot \ln n \cdot E_n(\partial_0 u), \\ \||z_y\|| &\leq cn^{-1/q} \cdot \ln n \cdot E_n(\partial_0 u), \end{aligned} \tag{3.395}$$

   where u is the solution of (3.349);

2. *the first equation in (3.389) can be written in the form*

$$x = G_1 x + S^{-1} Q(f_x + \tilde{f}_x) \tag{3.396}$$

   and can be solved by the fixed point iteration

$$x^{(k+1)} = G_1 x^{(k)} + S^{-1} Q(f_x + \tilde{f}_x), \quad k = 0, 1, \dots; \quad x^{(0)} \text{ is arbitrary} \tag{3.397}$$

   with the convergence rate of a geometrical progression with the denominator $q \leq c\frac{\ln n}{n^{1/q}} < 1$ *for n large enough.*

Proof. Using (B6) the norm of ψ_y can be estimated in the following way:

$$\begin{aligned} \||\psi_y\|| &= \max_{1 \leq k \leq n} \left| \int_{t_{k-1}}^{t_k} A_k^{(2)} e^{-A_k^{(2)}(t_k - \eta)} B_k^{(1)} [\mu_k - \mu(\eta)] [\theta(\eta) - P_n(\eta; \theta)] d\eta \right| \\ &\leq c \|\theta(\cdot) - P_n(\cdot; \theta)\|_{C[-1,1]} \max_{1 \leq k \leq n} (t_k - t_{k-1})^{1+1/q} \\ &\leq cn^{-(1+1/q)}(1 + \Lambda_n) E_n(\theta) \leq cn^{-(1+1/q)} \cdot \ln n \cdot E_n(\theta). \end{aligned}$$

Now the first assertion of the theorem follows from (3.393) and (3.394) and the second one from (3.392). □

Remark 3.55. Assuming that the vector-valued functions $\partial_0 u(t)$ can be analytically extended from the interval $[-1, 1]$ into the domain $\mathcal{D}_\rho$ enveloped by the Bernstein regularity ellipse $\mathcal{E}_\rho = \mathcal{E}_\rho(B)$ with the foci at $z = \pm 1$ and the sum of semiaxes equal to $\rho > 1$ (see also section 2.1), we obtain for the value of the best polynomial approximation [7, 27]

$$E_n(\partial_0 u) \leq \frac{\rho^{-n}}{1 - \rho} \sup_{z \in \mathcal{D}_\rho} \|\partial_0 u(z)\|,$$

which together with (3.395) yields the exponential convergence rate with the estimates (3.395).

Remark 3.56. For an elliptic operator A and for its discrete counterpart A_h (say, A_h is the finite element method/finite difference stiffness matrix corresponding to A), it holds that (see, e.g., [14])

$$\|(zI - A)^{-1}\|_{X \to X} \leq \frac{1}{|z| \sin (\theta_1 - \theta)} \qquad \forall z \in \mathbb{C} : \ \theta_1 \leq |\arg z| \leq \pi \qquad (3.398)$$

for any $\theta_1 \in (\theta, \pi)$, where $\cos \theta = \delta_0/C$, i.e., both are strongly positive in the sense of (3.323). Replacing A by A_h in (3.349) and then applying the temporal discretization described above, we arrive at the full discretization of problem (3.317). Since the bound (3.398) on the matrix resolvent of A_h is valid uniformly in the mesh size h, the full error is the sum of the temporal and the spatial errors which can be controlled independently from each other.

3.4.6   Examples

In this subsection, we show that many applied parabolic problems with a time-dependent boundary condition can be fitted into our abstract framework. Our aim here is to illustrate the assumptions accepted in sections 3.4.3 and 3.4.4.

A special example of the problem from the class (3.309) is

$$\frac{\partial u}{\partial t} = \frac{\partial^2 u}{\partial x^2} + f(x, t),$$
$$u(0, t) = 0, \qquad \frac{\partial u(1, t)}{\partial x} + b(t) u(1, t) = g(t), \qquad (3.399)$$
$$u(x, 0) = u_0(x),$$

where the operator $A : D(A) \in X \to X$, $X = L_q(0, 1)$, is defined by

$$D(A) = \{v \in W_q^2(0, 1) : v(0) = 0\},$$
$$Av = -\frac{\partial^2 v}{\partial x^2}, \qquad (3.400)$$

the operators $\partial_1 : D(A) \to Y$ and $\partial_0(t) : D(A) \to Y$, $Y = \mathbb{R}$ are defined by

$$\partial_1 u = \left. \frac{\partial u(x, t)}{\partial x} \right|_{x=1}, \qquad (3.401)$$
$$\partial_0(t) u = b(t) \cdot u(x, t)|_{x=1},$$

and $g(t) \in L_q(0, T; Y) = L_q(0, T)$.

We represent the solution of (3.399) in the form (compare with (3.319), (3.320))

$$u(x, t) = w(x, t) + v(x, t), \qquad (3.402)$$

where the function $v(x,t)$ is the solution of the problem

$$\frac{\partial v}{\partial t} = \frac{\partial^2 v}{\partial x^2} + f(x,t),$$

$$v(0,t) = 0, \qquad \frac{\partial v(1,t)}{\partial x} = 0, \tag{3.403}$$

$$v(x,0) = u_0(x)$$

and the function $w(x,t)$ satisfies

$$\frac{\partial w}{\partial t} = \frac{\partial^2 w}{\partial x^2},$$

$$w(0,t) = 0, \qquad \frac{\partial w(1,t)}{\partial x} = -b(t)u(1,t) + g(t), \tag{3.404}$$

$$w(x,0) = 0.$$

Introducing the operator $A^{(1)} : D(A^{(1)}) \to X$ defined by

$$D(A^{(1)}) = \{u \in W_q^2(0,1) : \ u(0) = 0, \ u'(1) = 0\} = \{u \in D(A) : u'(1) = 0\},$$

$$A^{(1)}u = -\frac{d^2u}{dx^2} \quad \forall u \in D(A^{(1)}) \tag{3.405}$$

(see (3.321) for an abstract setting), we can also write problem (3.403) in the form

$$\frac{dv}{dt} + A^{(1)}v = f(t),$$

$$v(0) = u_0, \tag{3.406}$$

with the solution

$$v(t) = e^{-A^{(1)}t}u_0 + \int_0^t e^{-A^{(1)}(t-\tau)} f(\tau)d\tau. \tag{3.407}$$

To solve the problem (3.404) we use the Duhamel integral. We introduce the auxiliary function $W(x,\lambda,t)$ satisfying the following equations (compare with (3.326)):

$$\frac{\partial W(x,\lambda,t)}{\partial t} = \frac{\partial^2 W(x,\lambda,t)}{\partial x^2},$$

$$W(0,\lambda,t) = 0, \qquad \frac{\partial W(1,\lambda,t)}{\partial x} = -b(\lambda)u(1,\lambda) + g(\lambda), \tag{3.408}$$

$$W(x,\lambda,0) = 0$$

with time-independent boundary conditions. Then the solution of problem (3.404) is given by (see Theorem 3.45)

$$w(x,t) = \int_0^t \frac{\partial}{\partial t} W(x,\lambda,t-\lambda)d\lambda. \tag{3.409}$$

Using this representation one can obtain an integral equation with respect to $u(1, t)$. Actually, we have

$$
\begin{aligned}
w(1, t) = u(1, t) - v(1, t) &= \frac{\partial}{\partial t} \int_0^t W(1, \lambda, t - \lambda) d\lambda \\
&= \int_0^t \frac{\partial}{\partial t} W(1, \lambda, t - \lambda) d\lambda.
\end{aligned}
\tag{3.410}
$$

The substitution

$$
W(x, \lambda, t) = B[-b(\lambda)u(1, \lambda) + g(\lambda)] + W_1(x, \lambda, t)
\tag{3.411}
$$

implies the following problem with homogeneous boundary conditions for $W_1(x, \lambda, t)$:

$$
\begin{aligned}
\frac{\partial W_1(x, \lambda, t)}{\partial t} &= \frac{\partial^2 W_1(x, \lambda, t)}{\partial x^2}, \\
W_1(0, \lambda, t) = 0, \quad & \frac{\partial W_1(1, \lambda, t)}{\partial x} = 0, \\
W_1(x, \lambda, 0) &= -B[-b(\lambda)u(1, \lambda) + g(\lambda)],
\end{aligned}
\tag{3.412}
$$

where the operator B is given by

$$
B[-b(\lambda)u(1, \lambda) + g(\lambda)] = x[-b(\lambda)u(1, \lambda) + g(\lambda)].
\tag{3.413}
$$

By separation of variables we get the solution of this problem explicitly in the form

$$
W_1(x, \lambda, t) = -2[-b(\lambda)u(1, \lambda) + g(\lambda)]
\tag{3.414}
$$

$$
\times \sum_{n=1}^{\infty} (-1)^{n+1} \left[\frac{2}{(2n-1)\pi} \right]^2 e^{-[\pi(2n-1)/2]^2 t} \sin \left[\frac{\pi}{2}(2n-1)x \right].
$$

Due to (3.413) the boundary integral equation (3.336) (or, equivalently, (3.339)) for the example problem (3.399) takes the form

$$
b(t)u(1, t) = b(t)v(1, t) + b(t) \int_0^t \frac{\partial}{\partial t} W_1(1, \lambda, t - \lambda) d\lambda
\tag{3.415}
$$

$$
= b(t)v(1, t) - b(t) \int_0^t K(t - \lambda)g(\lambda) d\lambda + b(t) \int_0^t K(t - \lambda)b(\lambda)u(1, \lambda) d\lambda,
$$

with

$$
K(t) = 2 \sum_{n=1}^{\infty} e^{-[\pi(2n-1)/2]^2 t}.
\tag{3.416}
$$

Remark 3.57. Note that in this particular case we can represent the integrand through the kernel $K(t - \lambda)$ analytically. In the general case, one can use the exponentially convergent algorithm for the operator exponential in (3.337) like the one from subsection 3.1.4.

Let us illustrate Theorem 3.47 for the example problem (3.399). It is easy to see that the condition (A1) is fulfilled provided that

$$|b(t)| \le c \quad \forall t \in [0, T]. \tag{3.417}$$

Let us show that there exists p for which the assumption (A2) holds. Actually, we have

$$\int_0^t \|A^{(1)} e^{-A^{(1)}(t-\lambda)} B\|_{Y \to X}^p d\lambda = \int_0^t [K(t-\lambda)]^p d\lambda$$

$$= \int_0^t \left[2 \sum_{n=1}^\infty e^{-(\frac{\pi(2n-1)}{2})^2 (t-\lambda)} \right]^p d\lambda. \tag{3.418}$$

Using the Hölder inequality, the kernel $K(t - \lambda)$ can be estimated by

$$K(t - \lambda) = 2 \sum_{n=1}^\infty e^{-(\frac{\pi(2n-1)}{2})^2 (t-\lambda)}$$

$$= 2 \sum_{n=1}^\infty (2n-1)^\alpha e^{-(\frac{\pi(2n-1)}{2})^2 (t-\lambda)} \frac{1}{(2n-1)^\alpha} \tag{3.419}$$

$$\le 2 \left\{ \sum_{n=1}^\infty (2n-1)^{p\alpha} e^{-p(\frac{\pi(2n-1)}{2})^2 (t-\lambda)} \right\}^{1/p} \cdot \left\{ \sum_{n=1}^\infty \frac{1}{(2n-1)^{q\alpha}} \right\}^{1/q},$$

where $\frac{1}{p} + \frac{1}{q} = 1$.

Substituting this inequality into (3.418), we obtain

$$\int_0^t [K(t-\lambda)]^p d\lambda \le \frac{2^{p+1}}{p\pi^2} \left\{ \sum_{n=1}^\infty \frac{1}{(2n-1)^{q\alpha}} \right\}^{p/q}$$

$$\times \sum_{n=1}^\infty \frac{1}{(2n-1)^{2-p\alpha}} \left[1 - e^{-p(\frac{\pi(2n-1)}{2})^2 t} \right] \tag{3.420}$$

$$\le \frac{2^{p+1}}{p\pi^2} \left\{ \sum_{n=1}^\infty \frac{1}{(2n-1)^{q\alpha}} \right\}^{p/q} \cdot \sum_{n=1}^\infty \frac{1}{(2n-1)^{2-p\alpha}}.$$

The series are bounded by a constant c if

$$\begin{cases} q\alpha > 1, \\ 2 - p\alpha > 1, \end{cases} \tag{3.421}$$

or $\frac{1}{q} < \alpha < \frac{1}{p}$. Thus, we can choose an arbitrary $p \in [1,2)$, and then $\frac{1}{q} = 1 - \frac{1}{p}$ and the choice of an arbitrary α from the interval $(1 - \frac{1}{p}, \frac{1}{p})$ provide that the assumption (A2) holds.

The assumption (A3) for the example problem (3.399) reads as

$$\left\{ \int_0^T \left| b(t)v(1,t) - b(t) \int_0^t K(t-\lambda)g(\lambda)d\lambda \right|^q dt \right\}^{1/q} \le c. \qquad (3.422)$$

Since $K(t-\cdot) \in L_p(0,T)$, this inequality holds provided that

$$g(t) \in L_q(0,T), \quad v(1,t) \in L_q(0,T). \qquad (3.423)$$

Due to the estimate

$$\left(\int_0^T |v(1,t)|^q dt \right)^{1/q} = \left(\int_0^T \left| \int_0^1 \frac{\partial v(x,t)}{\partial x} dx \right|^q dt \right)^{1/q}$$

$$\le \left(\int_0^T \int_0^1 \left| \frac{\partial v(x,t)}{\partial x} \right|^q dxdt \right)^{1/q}, \qquad (3.424)$$

the condition $v(1,t) \in L_q(0,T)$ is fulfilled if the last integral exists. The corresponding sufficient conditions on the input data $f(x,t)$, $u_0(x)$ of the problem (3.399) can be found in [54, 40].

Given the solution of (3.415) (which can be found numerically by algorithm (3.370) consisting in this case of the second equation only), the problem (3.399) takes the form

$$\frac{du}{dt} + Au = f(t),$$
$$u(0) = u_0, \quad \partial_1 u = g_1(t), \qquad (3.425)$$

with a known function $g_1(t) = g(t) - \partial_0(t)u(t)$. Using the representation by Duhamel's integral (3.409) we have

$$u(t) = v(t) - \int_0^t A^{(1)} e^{-A^{(1)}(t-\lambda)} Bg_1(\lambda)d\lambda. \qquad (3.426)$$

Now let us show that the assumptions (B1)–(B6) hold for the following model problem from the class (3.309):

$$\frac{\partial u(x,t)}{\partial t} = \frac{\partial^2 u}{\partial x^2} - q(x,t)u + f(x,t),$$
$$u(0,t) = 0, \quad \frac{\partial u(1,t)}{\partial x} + b(t)u(1,t) = g(t), \qquad (3.427)$$
$$u(x,0) = u_0(x).$$

The results of [40, section 4, paragraph 9, p. 388] yield that this problem possesses the unique solution $u \in W_q^{2,1}(Q_T)$ provided that $q(x,t)$, $f(x,t) \in L_q(Q_T)$, $Q_T = (0,1) \times (0,T)$, $g(t) \in W_q^{1-1/(2q)}(0,T)$, $b(t) \in W_q^{1/2}(0,T)$, $u_0 \in W_q^{2-\frac{2}{q}}(0,1)$, $q > 3/2$, and some compatibility conditions for initial and boundary conditions are fulfilled.

Here the operator $A(t) : D(A) \in X \to X$, $X = L_q(0,1)$, is defined by

$$D(A(t)) = \{v \in W_q^2(0,1) : v(0) = 0\},$$
$$A(t)v = -\frac{d^2v(x)}{dx^2} + q(x,t)v(x),$$

(3.428)

the operators $\partial_1 : D(A) \to Y$ and $\partial_0(t) : D(A) \to Y$, $Y = \mathbb{R}$ are defined by

$$\partial_1 u = \frac{\partial u(x,t)}{\partial x}\Big|_{x=1},$$
$$\partial_0(t)u = b(t) \cdot u(x,t)|_{x=1},$$

(3.429)

and $g(t) \in L_q(0,T;Y) = L_q(0,T)$. We suppose that the function $b(t)$ satisfies condition (3.417), so that our assumption (B5) is fulfilled.

The piecewise constant operator $A^{(2)}(t)$ from (3.354) is defined on each subinterval $[t_{k-1}, t_k]$ by

$$D(A^{(2)}(t)) = \{v \in W_q^2(0,1) : v(0) = 0, \quad v'(1) + \mu_k v(1) = 0\}, \quad \mu_k = b(t_k),$$
$$A_k v = -\frac{d^2v(x)}{dx^2} + q(x,t_k)v(x)$$

and the operator $B_k^{(1)}$ by

$$B_k^{(1)} z = \frac{x}{\mu_k + 1} z, \quad z \in Y.$$

Assumptions (B1)–(B2) are fulfilled due to results from [14], and assumptions (B3)–(B4) are obviously fulfilled, too. It remains to check assumption (B6). Let $(\lambda_j^{(k)}, e_j^{(k)})$, $j = 1, 2, \ldots$, be the system of eigenpairs of $A^{(2)}(t)$, $t \in [t_{k-1}, t_k]$. Then we have

$$\int_0^t \|[A^{(2)}]^{1+\gamma} e^{-A^{(2)}(t-\lambda)} B^{(1)}\|_{Y \to X}^p d\lambda = \sum_{k=1}^n \int_{t_{k-1}}^{t_k} \left[K_\gamma^{(k)}(t-\lambda)\right]^p d\lambda, \quad (3.430)$$

where

$$K_\gamma^{(k)}(t-\lambda) = \sum_{j=1}^\infty [\lambda_j^{(k)}]^\gamma e^{-\lambda_j^{(k)}(t-\lambda)}. \quad (3.431)$$

Using the well-known asymptotic $c_L j^2 \leq \lambda_j \leq c_U j^2$ (see, e.g, [45]) and Hölder's inequality, the kernel $K_\gamma^{(k)}(t-\lambda)$ can be estimated in the following way:

$$
K_\gamma^{(k)}(t-\lambda) = \sum_{j=1}^{\infty} \left(\lambda_j^{(k)}\right)^\gamma e^{-\lambda_j^{(k)}(t-\lambda)} \leq c_U^\gamma \sum_{j=1}^{\infty} j^{2\gamma} e^{-c_L j^2 (t-\lambda)}
$$

$$
= c_U^\gamma \sum_{n=1}^{\infty} j^{2\gamma+\alpha} e^{-c_L j^2 (t-\lambda)} \frac{1}{j^\alpha}
$$

$$
\leq c_U^\gamma \left\{ \sum_{j=1}^{\infty} j^{p(\alpha+2\gamma)} e^{-pc_L j^2 (t-\lambda)} \right\}^{1/p} \cdot \left\{ \sum_{j=1}^{\infty} \frac{1}{j^{q\alpha}} \right\}^{1/q}, \quad \frac{1}{p} + \frac{1}{q} = 1.
$$

Substituting this inequality into (3.430) we obtain

$$
\sum_{k=1}^{n} \int_{t_{k-1}}^{t_k} [K_\gamma^{(k)}(t-\lambda)]^p d\lambda
$$

$$
\leq \sum_{k=1}^{n} c_U^\gamma \int_{t_{k-1}}^{t_k} \sum_{j=1}^{\infty} j^{p(\alpha+2\gamma)} e^{-pc_L j^2 (t-\lambda)} d\lambda \cdot \left\{ \sum_{j=1}^{\infty} \frac{1}{j^{q\alpha}} \right\}^{p/q}
$$

$$
= c_U^\gamma \sum_{j=1}^{\infty} j^{p(\alpha+2\gamma)} (pc_L j^2)^{-1} \left[1 - e^{-pc_L j^2 t} \right] \cdot \left\{ \sum_{j=1}^{\infty} \frac{1}{j^{q\alpha}} \right\}^{p/q}
$$

$$
\leq \frac{c_U^\gamma}{pc_L} \sum_{j=1}^{\infty} \frac{1}{j^{2-p(\alpha+2\gamma)}} \cdot \left\{ \sum_{j=1}^{\infty} \frac{1}{j^{q\alpha}} \right\}^{p/q}.
$$

The series remains bounded if

$$
\begin{cases} q\alpha > 1, \\ 2 - p(\alpha + 2\gamma) > 1, \end{cases}
$$

or $\frac{1}{q} < \alpha < \frac{1}{p} - 2\gamma$. Thus, we can choose an arbitrary $p \in [1,2)$; then $\frac{1}{q} = 1 - \frac{1}{p}$ and the assumption (B6) holds with an arbitrary γ from the interval $[0, \frac{1}{p} - \frac{1}{2})$.

3.4.7   Numerical example

In this subsection we show that the algorithm (3.370) possesses the exponential convergence with respect to the temporal discretization parameter n predicted by Theorem 3.54. To eliminate the influence of other errors (the spatial error, the error of approximation of the operator exponential and of the integrals in (3.371)), we calculate the coefficients of the algorithm (3.372) exactly using the computer algebra tool Maple.

We consider the model problem (3.399) with

$$u_0(x) = \frac{4\sqrt{2}}{\pi} \sin \frac{\pi}{4} x, \quad b(t) = \frac{\pi}{4} \exp\left(\frac{\pi^2}{16} t\right), \quad g(t) = 1 + \exp\left(-\frac{\pi^2}{16} t\right),$$

for which algorithm (3.370) consists of the second equation only, where $\alpha_{kj} = 0$. The exact solution of this problem is

$$u(x,t) = \exp\left(-\frac{\pi^2}{16} t\right) \frac{4\sqrt{2}}{\pi} \sin \frac{\pi}{4} x.$$

For the function $v(x,t)$ from (3.403) we have the following problem:

$$\frac{\partial v(x,t)}{\partial t} = \frac{\partial^2 v(x,t)}{\partial x^2}, \quad v(0,t) = 0, \quad \frac{\partial v(1,t)}{\partial x} = 0,$$

$$v(x,0) = \frac{4\sqrt{2}}{\pi} \sin \frac{\pi}{4} x.$$

It is easy to check that

$$v(x,t) = \sum_{k=1}^{\infty} \frac{(-1)^{k+1} 32}{\pi^2 (4k-1)(4k-3)} e^{-\left(\frac{\pi(2k-1)}{2}\right)^2 t} \sin\left(\frac{\pi}{2}(2k-1)x\right).$$

Equation (3.415) (compare also with (3.414), (3.420)) reads as follows:

$$u(1,t) = v(1,t) + 2 \int_0^t \left[1 + e^{\frac{-\pi^2}{16}\tau}\right] \sum_{k=1}^{\infty} e^{-\left(\frac{\pi(2k-1)}{2}\right)^2(t-\tau)} d\tau$$

$$- 2 \int_0^t \frac{\pi}{4} e^{\frac{\pi^2}{16}\tau} u(1,\tau) \sum_{k=1}^{\infty} e^{-\left(\frac{\pi(2k-1)}{2}\right)^2(t-\tau)} d\tau$$

$$= f(t) - \frac{\pi}{2} \int_0^t e^{\frac{\pi^2}{16}\tau} u(1,\tau) \sum_{k=1}^{\infty} e^{-\left(\frac{\pi(2k-1)}{2}\right)^2(t-\tau)} d\tau, \qquad (3.432)$$

with

$$f(t) = v(1,t) - 2 \sum_{k=1}^{\infty} e^{-\left(\frac{\pi(2k-1)}{2}\right)^2 t} \left[\frac{1}{\left(\frac{\pi(2k-1)}{2}\right)^2} + \frac{1}{\left(\frac{\pi(2k-1)}{2}\right)^2 - \frac{\pi^2}{16}}\right] + 1$$

$$+ \frac{4}{\pi} e^{-\frac{\pi^2}{16} t} = -\frac{8}{\pi^2} \sum_{k=1}^{\infty} \frac{e^{-\left(\frac{\pi(2k-1)}{2}\right)^2 t}}{(2k-1)^2} + 1 + \frac{4}{\pi} e^{-\frac{\pi^2}{16} t}.$$

To reduce the problem to the interval $[-1,1]$ we change the variable t in (3.432) by

$$t = \frac{s+1}{2} T, \quad s \in [-1,1]; \quad \tau = \frac{\xi+1}{2} T, \quad \xi \in [-1,s]; \quad d\tau = \frac{T}{2} d\xi$$

and obtain

$$u\left(1, \frac{s+1}{2}T\right) = f\left(\frac{s+1}{2}T\right)$$

$$-\frac{\pi T}{4}\sum_{k=1}^{\infty}\int_{-1}^{s} e^{\frac{\pi^2}{16}\frac{\xi+1}{2}T}u\left(1, \frac{\xi+1}{2}T\right)e^{-\left(\frac{\pi(2k-1)}{2}\right)^2\frac{T}{2}(s-\xi)}d\xi.$$

Note that this equation is, in fact, the explicit form of the operator equation (3.420) with an explicitly calculated integrand including the operator exponential.

Substituting the corresponding interpolation polynomials and collocating the obtained equation at the points

$$s_l = \cos\left[\frac{2l-1}{2n}\pi\right], \quad l = \overline{1, n},$$

analogously as in (3.372), we arrive at a system of linear algebraic equations with respect to $y_j \approx u(1, t_j)$, where the calculation of the values β_{kj} and $\partial_0\phi_k$ is reduced to the calculation of integrals of the type $\int_{-1}^{s}e^{-a(s-\tau)}\tau^n d\tau$ and can be performed exactly. The computations were provided in Maple 9. For the cases of $n = 2, 3, 4, 8$ we used Digits=20, whereas we used Digits=40 for the case $n = 16$. The results of the computation presented in Tables 3.9–3.12 confirm our theory above.

Point t	ε
0.8535533905	0.64328761e-2
0.1464466094	0.1817083e-2

Table 3.9: The error in the case $n = 2$, $T = 1$.

Point t	ε
0.9619397662	0.11818295e-4
0.6913417161	0.98439031e-5
0.3086582838	0.71990794e-5
0.0380602337	0.18751078e-5

Table 3.10: The error in the case $n = 4$, $T = 1$.

Point t	ε
0.9903926402	0.23186560e-11
0.9157348061	0.26419501e-11
0.7777851165	0.2614635e-11
0.5975451610	0.2367290e-11
0.4024548389	0.2059591e-11
0.2222148834	0.1600175e-11
0.0842651938	0.1298227e-11
0.0096073597	0.3285404e-12

Table 3.11: The error in the case $n = 8$, $T = 1$.

Point t	ε
0.9975923633	0.25187858260140e-26
0.9784701678	0.17507671148106e-26
0.9409606321	0.26606504614832e-26
0.8865052266	0.19216415631333e-26
0.8171966420	0.14389447843978e-26
0.7356983684	0.8729247130003e-27
0.6451423386	0.12250410788313e-26
0.5490085701	0.20057933602212e-26
0.4509914298	0.27443715793522e-26
0.3548576613	0.2071356401636e-26
0.2643016315	0.1352407675441e-26
0.1828033579	0.4176006046401e-26
0.1134947733	0.5890397288154e-26
0.0590393678	0.49643071858013e-25
0.0215298321	0.4614912514700e-26
0.0024076366	0.12497053453746e-25

Table 3.12: The error in the case $n = 16$, $T = 1$.

Chapter 4

The second-order equations

This chapter is devoted to studying the problems associated with second-order differential equations with an unbounded operator coefficient A in a Banach space. In Section 4.1, we consider these equations with an unbounded operator in either Banach or Hilbert spaces depending on the parameter t. We propose a discretization method with a high parallelism level and without accuracy saturation, i.e., the accuracy adapts automatically to the smoothness of the solution. For analytical solutions, the rate of convergence is exponential. These results can be viewed as a development of parallel approximations of the operator cosine function $\cos(\sqrt{A}t)$ which represents the solution operator of the initial value problem for the second-order differential equation with a constant operator A coefficient.

Section 4.2 focuses on the second-order strongly damped differential equation with operator coefficients in a Banach space. A new fast convergent algorithm is proposed. This algorithm is based on the Dunford–Cauchy integral representation and on the Sinc-quadratures providing an exponential convergence rate of the algorithm. Examples of initially-boundary value problems for the strongly damped wave equation are given that justify the theoretical results.

In section 4.3, we consider the second-order differential equation with an unbounded operator coefficient in a Banach space equipped with a boundary condition which can be viewed as a meta-model for elliptic PDEs. The solution can be written down by using the normalized operator hyperbolic sine family $\sinh^{-1}(\sqrt{A})\sinh(x\sqrt{A})$ as a solution operator. The solution of the corresponding inhomogeneous boundary value problem is found through a solution operator and a Green function. Starting with the Dunford-Cauchy representation for a normalized hyperbolic operator sine family and for a Green function, we use discretization of the integrals involving the exponentially convergent Sinc quadratures which leads to a short sum of resolvents of A. Our algorithm inherits a two-level parallelism with respect to both the computation of resolvents and the evaluation for different values of the spatial variable $x \in [0, 1]$.

4.1 Algorithm without accuracy saturation for second-order evolution equations in Hilbert and Banach spaces

4.1.1 Introduction

In this section, we consider the evolution problem

$$\frac{d^2u}{dt^2} + A(t)u = f(t), \quad t \in (0,T]; \quad u(0) = u_0, \quad u'(0) = u_{01} \tag{4.1}$$

where $A(t)$ is a densely defined closed (unbounded) operator with domain $D(A)$ independent of t in a Banach space X, u_0, u_{01} are given vectors, and $f(t)$ is a given vector-valued function. We suppose the operator $A(t)$ to be strongly positive; i.e., there exists a positive constant M_R independent of t such that on the rays and outside a sector $\Sigma_\theta = \{z \in \mathbb{C} : 0 \leq arg(z) \leq \theta, \ \theta \in (0, \pi/2)\}$ the following resolvent estimate holds:

$$\|(zI - A(t))^{-1}\| \leq \frac{M_R}{1 + |z|}. \tag{4.2}$$

This assumption implies that there exists a positive constant c_κ such that (see [14], p. 103)

$$\|A^\kappa(t)e^{-sA(t)}\| \leq c_\kappa s^{-\kappa}, \quad s > 0, \quad \kappa \geq 0. \tag{4.3}$$

Our further assumption is that there exists a real positive ω such that

$$\|e^{-sA(t)}\| \leq e^{-\omega s} \quad \forall s, \ t \in [0,T] \tag{4.4}$$

(see [54], Corollary 3.8, p. 12, for corresponding assumptions on $A(t)$). Let us also assume that the conditions

$$\|[A(t) - A(s)]A^{-\gamma}(t)\| \leq \tilde{L}_{1,\gamma}|t - s| \quad \forall t, \ s, \ 0 \leq \gamma \leq 1, \tag{4.5}$$

$$\|A^\beta(t)A^{-\beta}(s) - I\| \leq \tilde{L}_\beta|t - s| \quad \forall t, \ s \in [0,T] \tag{4.6}$$

hold.

The example is considered in subsection 3.2.1 which shows the practical relevance for the assumptions above .

4.1.2 Discrete second-order problem

In this subsection, we consider the problem (4.1) in a Hilbert space H with the scalar product $(\cdot, \cdot)$ and the corresponding norm $\| \cdot \| = \sqrt{(\cdot, \cdot)}$. Let assumption (4.6) related to $A(t)$ hold. In addition, we assume that the assumption (3.121) holds. Namely:

$$A(t) = \sum_{k=0}^{m_A} A_k t^k, \quad f(t) = \sum_{k=0}^{m_f} f_k t^k \tag{4.7}$$

and

$$\|A^{\gamma-1/2}(0)(A(t) - A(s))A^{-\gamma}(0)\| \leq \tilde{L}_{2,\gamma}|t - s|^{\alpha}, \gamma, \alpha \in [0, 1]. \qquad (4.8)$$

Remark 4.1. Condition (4.8) coincides with (4.5) for $\gamma = 1/2$, $\alpha = 1$.

Let $\overline{A}(t)$ be a piece-wise constant operator defined on the grid ω_n of n Chebyshev points as in subsection 3.2.2,

$$\overline{A}(t) = A_k = A(t_k), \quad t \in (t_{k-1}, t_k],$$

We consider the auxiliary problem

$$\frac{d^2u}{dt^2} + \overline{A}(t)u = [\overline{A}(t) - A(t)] + f(t), \qquad (4.9)$$
$$u(-1) = u_0, \quad u'(1) = u'_0,$$

from which we get the following relations for the interval $[t_{k-1}, t_k]$:

$$u(t) = \cos\left[\sqrt{A_k}(t - t_{k-1})\right]u(t_{t_{k-1}}) + A_k^{-1/2}\sin\left[\sqrt{A_k}(t - t_{k-1})\right]u'(t_{t_{k-1}})$$
$$+ \int_{t_{k-1}}^{t_k} A_k^{-1/2}\sin\left[\sqrt{A_k}(t - \eta)\right]\{[\overline{A}(\eta) - A(\eta)]u(\eta) + f(\eta)\}d\eta,$$
$$u'(t) = -\sqrt{A_k}\sin\left[\sqrt{A_k}(t - t_{k-1})\right]u(t_{t_{k-1}}) + \cos\left[\sqrt{A_k}(t - t_{k-1})\right]u'(t_{t_{k-1}}) \qquad (4.10)$$
$$+ \int_{t_{k-1}}^{t_k} \cos\left[\sqrt{A_k}(t - \eta)\right]\{[\overline{A}(\eta) - A(\eta)]u(\eta) + f(\eta)\}d\eta.$$

Let

$$P_{n-1}(t; u) = P_{n-1}u = \sum_{p=1}^{n} u(t_p)L_{p,n-1}(t)$$

be the interpolation polynomial for the function $u(t)$ on the mesh ω_n, let $y = (y_1, \ldots, y_n)$, $y_i \in X$ be a given vector, and let

$$P_{n-1}(t; y) = P_{n-1}y = \sum_{p=1}^{n} y_p L_{p,n-1}(t) \qquad (4.11)$$

be the polynomial that interpolates y, where $L_{p,n-1} = \frac{T_n(t)}{T'_n(t_p)(t-t_p)}$, $p = 1, \ldots, n$, are the Lagrange fundamental polynomials.

We substitute the interpolation polynomial (4.11) in (4.10) instead of u. Then by collocation, we arrive at the following system of linear algebraic equations with respect to unknowns y_k, y'_k which approximate $u(t_k)$ and $u'(t_k)$, respectively, i.e.,

$$y_k = \cos\left[\sqrt{A_k}\,\tau_k\right]y_{k-1} + A_k^{-1/2}\sin\left[\sqrt{A_k}\,\tau_k\right]y'_{k-1}$$
$$+ \sum_{i=1}^{n}\alpha_{k,i}y_i + \phi_k^{(1)},$$
$$y'_k = -\sqrt{A_k}\sin\left[\sqrt{A_k}\,\tau_k\right]y_{k-1} + \cos\left[\sqrt{A_k}\,\tau_k\right]y'_{k-1} \qquad (4.12)$$
$$+ \sum_{i=1}^{n}\beta_{k,i}y_i + \phi_k^{(2)}, \quad k = 1,2,\ldots,n,$$
$$y_0 = u_0, \quad y'_0 = u'_0,$$

where

$$\alpha_{k,i} = \int_{t_{k-1}}^{t_k} A_k^{-1/2}\sin\left[\sqrt{A_k}(t_k - \eta)\right][A_k - A(\eta)]L_{i,n-1}(\eta)d\eta,$$
$$\beta_{k,i} = \int_{t_{k-1}}^{t_k} \cos\left[\sqrt{A_k}(t_k - \eta)\right][A_k - A(\eta)]L_{i,n-1}(\eta)d\eta,$$
$$\phi_k^{(1)} = \int_{t_{k-1}}^{t_k} A_k^{-1/2}\sin\left[\sqrt{A_k}(t_k - \eta)\right]f(\eta)d\eta, \qquad (4.13)$$
$$\phi_k^{(2)} = \int_{t_{k-1}}^{t_k} \cos\left[\sqrt{A_k}(t_k - \eta)\right]f(\eta)d\eta, \quad k = 1,2,\ldots,n.$$

The errors $z_k = u(t_k) - y_k$, $z'_k = u'(t_k) - y'_k$ satisfy the equations

$$z_k = \cos\left[\sqrt{A_k}\,\tau_k\right]z_{k-1} + A_k^{-1/2}\sin\left[\sqrt{A_k}\,\tau_k\right]z'_{k-1}$$
$$+ \sum_{i=1}^{n}\alpha_{k,i}z_i + \psi_k^{(1)},$$
$$z'_k = -\sqrt{A_k}\sin\left[\sqrt{A_k}\,\tau_k\right]z_{k-1} + \cos\left[\sqrt{A_k}\,\tau_k\right]z'_{k-1} \qquad (4.14)$$
$$+ \sum_{i=1}^{n}\beta_{k,i}z_i + \psi_k^{(2)}, \quad k = 1,2,\ldots,n,$$
$$z_0 = 0, \quad z'_0 = 0,$$

where

$$\psi_k^{(1)} = \int_{t_{k-1}}^{t_k} A_k^{-1/2}\sin\left[\sqrt{A_k}(t_k - \eta)\right][A_k - A(\eta)][u(\eta) - P_{n-1}(\eta;u)]d\eta,$$
$$\psi_k^{(2)} = \int_{t_{k-1}}^{t_k} \cos\left[\sqrt{A_k}(t_k - \eta)\right][A_k - A(\eta)][u(\eta) - P_{n-1}(\eta;u)]d\eta, \qquad (4.15)$$
$$k = 1,2,\ldots,n.$$

Let us write $\tilde{y}_k = A_k^\gamma y_k$, $\tilde{y}_k' = A_k^{\gamma-1/2} y_k'$, $\tilde{z}_k = A_k^\gamma z_k$, $\tilde{z}_k' = A_k^{\gamma-1/2} z_k'$ and rewrite (4.14) in the form

$$
\tilde{z}_k = \cos\left[\sqrt{A_k}\tau_k\right](A_k^\gamma A_{k-1}^{-\gamma})\tilde{z}_{k-1} + \sin\left[\sqrt{A_k}\tau_k\right](A_k^{\gamma-1/2} A_{k-1}^{-(\gamma-1/2)})\tilde{z}_{k-1}'
$$

$$
+ \sum_{i=1}^{n} \tilde{\alpha}_{k,i}\tilde{z}_i + \tilde{\psi}_k^{(1)},
$$

$$
\tilde{z}_k' = -\sin\left[\sqrt{A_k}\tau_k\right](A_k^\gamma A_{k-1}^{-\gamma})\tilde{z}_{k-1} + \cos\left[\sqrt{A_k}\tau_k\right](A_k^{\gamma-1/2} A_{k-1}^{-(\gamma-1/2)})\tilde{z}_{k-1}' \qquad (4.16)
$$

$$
+ \sum_{i=1}^{n} \tilde{\beta}_{k,i}\tilde{z}_i + \tilde{\psi}_k^{(2)}, \quad k = 1, 2, \ldots, n,
$$

$$
z_0 = 0, \quad \tilde{z}_0' = 0,
$$

where

$$
\tilde{\alpha}_{k,i} = A_k^\gamma \alpha_{k,i} A_i^{-\gamma}
$$
$$
= \int_{t_{k-1}}^{t_k} A_k^{\gamma-1/2} \sin\left[\sqrt{A_k}(t_k - \eta)\right][A_k - A(\eta)]A_i^{-\gamma} L_{i,n-1}(\eta)\,d\eta,
$$

$$
\tilde{\beta}_{k,i} = A_k^{\gamma-1/2} \beta_{k,i} A_i^{-\gamma}
$$
$$
= \int_{t_{k-1}}^{t_k} A_k^{\gamma-1/2} \cos\left[\sqrt{A_k}(t_k - \eta)\right][A_k - A(\eta)]A_i^{-\gamma} L_{i,n-1}(\eta)\,d\eta,
$$

$$
\tilde{\psi}_k^{(1)} = A_k^\gamma \phi_k^{(1)} \qquad\qquad (4.17)
$$
$$
= \int_{t_{k-1}}^{t_k} A_k^{\gamma-1/2} \sin\left[\sqrt{A_k}(t_k - \eta)\right] A_k^{-\gamma}[A_k^\gamma(u(\eta) - P_{n-1}(\eta; u))]\,d\eta,
$$

$$
\tilde{\psi}_k^{(2)} = A_k^{\gamma-1/2} \phi_k^{(2)}
$$
$$
= \int_{t_{k-1}}^{t_k} A_k^{\gamma-1/2} \cos\left[\sqrt{A_k}(t_k - \eta)\right] A_k^{-\gamma}[A_k^\gamma(u(\eta) - P_{n-1}(\eta; u))]\,d\eta.
$$

We introduce the 2×2 block matrices

$$
E = \begin{pmatrix} I & 0 \\ 0 & I \end{pmatrix},
$$

$$
B_k(t_k - \eta) = \begin{pmatrix} \cos(\sqrt{A_k})(t_k - \eta) & \sin(\sqrt{A_k})(t_k - \eta) \\ -\sin(\sqrt{A_k})(t_k - \eta) & \cos(\sqrt{A_k})(t_k - \eta) \end{pmatrix},
$$

$$
D_k = \begin{pmatrix} A_k^\gamma A_{k-1}^{-\gamma} & 0 \\ 0 & A_k^{\gamma-1/2} A_{k-1}^{-(\gamma-1/2)} \end{pmatrix},
$$

$$
F_i(\eta) = \begin{pmatrix} A_k^{\gamma-1/2}[A_i - A(\eta)]A_i^{-\gamma} & 0 \\ 0 & 0 \end{pmatrix},
$$

and

$$S = \begin{pmatrix} E & 0 & 0 & \cdots & 0 & 0 & 0 \\ -B_2 D_2 & E & 0 & \cdots & 0 & 0 & 0 \\ 0 & -B_3 D_3 & E & \cdots & 0 & 0 & 0 \\ \cdot & \cdot & \cdot & \cdots & \cdot & & \cdot \\ 0 & 0 & 0 & \cdots & -B_{n-1}D_{n-1} & E & 0 \\ 0 & 0 & 0 & \cdots & 0 & -B_n D_n & E \end{pmatrix},$$

$$C \equiv \{c_{i,j}\}_{i,j=1}^n = \begin{pmatrix} \tilde{\alpha}_{11} & 0 & \tilde{\alpha}_{12} & 0 & \cdots & \tilde{\alpha}_{1n} & 0 \\ \tilde{\beta}_{11} & 0 & \tilde{\beta}_{12} & 0 & \cdots & \tilde{\beta}_{1n} & 0 \\ \cdot & \cdot & \cdot & \cdot & \cdots & \cdot & \cdot \\ \tilde{\alpha}_{n1} & 0 & \tilde{\alpha}_{n2} & 0 & \cdots & \tilde{\alpha}_{nn} & 0 \\ \tilde{\beta}_{n1} & 0 & \tilde{\beta}_{n2} & 0 & \cdots & \tilde{\beta}_{nn} & 0 \end{pmatrix}$$

with $B_k = B_k(t_k - t_{k-1}) = B_k(\tau_k)$ and the 2×2 operator blocks

$$c_{i,j} = \begin{pmatrix} \tilde{\alpha}_{i,j} & 0 \\ \tilde{\beta}_{i,j} & 0 \end{pmatrix}.$$

These blocks can also be represented as

$$c_{i,j} = \int_{t_{i-1}}^{t_i} L_{j,n}(\eta) D_i^* B_i^* (t_i - \eta) F_i(\eta) d\eta.$$

Using the integral representation of functions of a self-adjoint operator by the corresponding spectral family, one can easily show that

$$B_k B_k^* = B_k^* B_k = E, B_k(t_k - \eta)B_k(t_k - \eta)^* = B_k(t_k - \eta)^* B_k(t_k - \eta) = E. \quad (4.18)$$

Analogously to the previous section, we get

$$S^{-1} \equiv \{s_{i,k}^{(-1)}\}_{i,k=1}^n$$
$$= \begin{pmatrix} E & 0 & 0 & \cdots & 0 \\ B_2 D_2 & E & 0 & \cdots & 0 \\ B_3 D_3 B_2 D_2 & B_3 D_3 & E & \cdots & 0 \\ \cdot & \cdot & \cdot & \cdots & \cdot \\ B_n D_n \cdots B_2 D_2 & B_n D_n \cdots B_3 D_3 & B_n D_n \cdots B_4 D_4 & \cdots & E \end{pmatrix} \quad (4.19)$$

with 2×2 operator block-elements $s_{i,k}^{(-1)}$. We introduce the vectors

$$z = (w_1, w_2, \ldots, w_n)$$
$$= (\tilde{z}_1, z_1', \ldots, \tilde{z}_n, \quad z_n'), \quad w_i = (z_i, z_i'), z_i, z_i' \in H,$$
$$\psi = (\Psi_1, \Psi_2, \ldots, \Psi_n)$$
$$= (\tilde{\psi}_1^{(1)}, \psi_1^{(2)}, \ldots, \tilde{\psi}_n^{(1)}, \psi_n^{(2)}), \quad \Psi_i = (\tilde{\psi}_i^{(i)}, \psi_i^{(2)}), \tilde{\psi}_i^{(i)}, \psi_i^{(2)} \in H.$$

Then equations (4.16) can be written in the block matrix form

$$z = S^{-1}Cz + S^{-1}\psi. \qquad (4.20)$$

Note that the vector-blocks Ψ_i can be written as

$$\Psi_i = \int_{t_{i-1}}^{t_i} \tilde{B}_i^*(t_i - \eta) D_\psi(\eta) d\eta \qquad (4.21)$$

with the block-vectors

$$D_\psi(\eta) = \begin{pmatrix} 0 \\ A^{\gamma-1/2}[A_i - A(\eta)]A_i^{-\gamma}[A_i^\gamma(u(\eta) - P_n(\eta; u))] \end{pmatrix}. \qquad (4.22)$$

The blocks E, B_i act in the space of two-dimensional block-vectors $v = (v_1, v_2)$, $v_1, v_2 \in H$. In this space, we define the new scalar product by

$$((u, v)) = (u_1, v_1) + (u_2, v_2), \qquad (4.23)$$

the corresponding block -vector norm by

$$|||v|||_b = \sqrt{((v, v))} = \left(\|v_1\|^2 + \|v_2\|^2\right)^{1/2} \qquad (4.24)$$

and the consistent norm for a block operator matrix $G = \begin{pmatrix} g_{11} & g_{12} \\ g_{21} & g_{22} \end{pmatrix}$ by

$$|||G|||_b = \sup_{y \neq 0} \frac{\sqrt{((Gv, Gv))}}{|||v|||}.$$

In the space of n-dimensional block-vectors, we define the block-vector norm by

$$|||y||| = \max_{1 \leq k \leq n} |||v_k|||_b \qquad (4.25)$$

and the consistent matrix norm

$$|||C||| \equiv |||\{c_{i,j}\}_{i,j=1}^n||| = \sup_{y \neq 0} \frac{|||Cy|||}{|||y|||} = \max_{1 \leq k \leq n} \sum_{p=1}^n |||c_{kp}|||_b. \qquad (4.26)$$

It is easy to see that

$$|||B_i|||_b = |||B_i^*|||_b = \sup_{v \neq 0} \frac{|||B_iv|||}{|||v|||} = \sup_{v \neq 0} \frac{\sqrt{((B_iv, B_iv))}}{|||v|||}$$

$$= \sup_{v \neq 0} \frac{\sqrt{((B_iB_i^*v, v))}}{|||v|||} = 1 \qquad (4.27)$$

and, due to (4.6),

$$
\begin{aligned}
|||D_k|||_b &= \sup_{v \neq 0} \frac{|||D_k v|||}{|||v|||} = \sup_{v \neq 0} \frac{\sqrt{((D_k v, D_k v))}}{|||v|||} \\
&= \sup_{v \neq 0} \frac{\sqrt{((D_k D_i k^* v, v))}}{|||v|||} \leq c_D, \\
|||D_k^*|||_b &\leq c_D
\end{aligned}
\tag{4.28}
$$

with $c_D = \sqrt{(1 + \tilde{L}_\gamma T)^2 + (1 + \tilde{L}_{\gamma-1/2} T)^2}$. Let us estimate $|||S^{-1}|||$. Due to (4.27), (4.28), (4.26) and (4.19) we have

$$
|||S^{-1}||| \leq c_D n.
\tag{4.29}
$$

Using assumptions (4.8) and (4.18), (4.29), we get

$$
\begin{aligned}
|||c_{i,j}|||_b &\leq \int_{t_{i-1}}^{t_i} |||D_i^*||| \, |||B_i(t_i - \eta)||| \, |||F_i(\eta)||| \, |L_{j,n}(\eta)| d\eta \\
&\leq c_D \tilde{L}_{2,\gamma} \tau_{\max}^\alpha \int_{t_{i-1}}^{t_i} |L_{j,n-1}(\eta)|^2 d\eta, \\
|||C||| &\leq \max_{1 \leq k \leq n} \sum_{p=1}^n |||c_{k,p}|||_b \\
&\leq c \tau_{\max}^\alpha \left(\int_{t_{k-1}}^{t_k} \sum_{j=1}^n |L_{j,n}(\eta)| d\eta \right) \leq c \Lambda_n \tau_{\max}^{1+\alpha} \leq c n^{-1-\alpha} \ln n, \\
|||S^{-1} C||| &\leq n^{-\alpha} \ln n
\end{aligned}
\tag{4.30}
$$

with an appropriate positive constant c independent of n.

Now we are in a position to prove the main result of this section.

Theorem 4.2. *Let assumptions* (4.2)–(4.7), (4.8), (4.6) *hold. Then there exists a positive constant c such that the following hold.*

1. *For n large enough it holds that*

$$
|||z||| \equiv |||y - u||| \leq c n^{-\alpha} \ln n \, E_n(A_0^\gamma u),
\tag{4.31}
$$

   where u is the solution of (4.1) *and $E_n(A_0^\gamma u)$ is the best approximation of $A_0^\gamma u$ by polynomials of degree not greater than $n - 1$.*

2. *The system of linear algebraic equations*

$$
Sy = Cy + f
\tag{4.32}
$$

with respect to the approximate solution y can be solved by the fixed-point iteration

$$y^{(k+1)} = S^{-1}Cy^{(k)} + S^{-1}f, \quad k = 0, 1, \ldots; \quad y^{(0)} \text{ arbitrary,} \qquad (4.33)$$

which converges as a geometric progression with the denominator

$$q = cn^{-\alpha}\ln n < 1$$

for n large enough.

Proof. Due to (4.20), (4.29) for $\tau_{\max}$ small enough (or for n large enough) there exists a bounded norm $|||(E - S^{-1}S)^{-1}|||$ and we get

$$|||z||| \le cn|||\psi|||. \qquad (4.34)$$

It remains to estimate $|||\psi|||$. Using (4.24), (4.21), (4.22) and (4.18), we have

$$\begin{aligned}
|||\psi||| &= \max_{1 \le k \le n} |||\Psi_k|||_b \\
&= \max_{1 \le k \le n} ||| \int_{t_{k-1}}^{t_k} \tilde{B}_k^*(t_k - \eta)D_\psi(\eta)d\eta||| \\
&\le \max_{1 \le k \le n} \int_{t_{k-1}}^{t_k} |t_k - \eta|^\alpha \|A_k^\gamma A_0^{-\gamma}\| \||[A_0^\gamma(u(\eta) - P_n(\eta; u))]\|d\eta \\
&\le (1 + \tilde{L}_\gamma T)\tau_{\max}^{1+\alpha}(1 + \Lambda_n)E_n(A_0^\gamma u) \\
&\le c(1 + \tilde{L}_\gamma T)n^{-1-\alpha}\ln n E_n(A_0^\gamma u).
\end{aligned} \qquad (4.35)$$

This inequality together with (4.34) completes the proof of the first assertion. The second one can be proved analogously to Theorem 3.25. $\qquad \square$

Remark 4.3. We arrive at an exponential accuracy for piecewise analytical solutions if we apply the methods, described above, successively on each subinterval of the analyticity.

4.2 Exponentially convergent algorithm for a strongly damped differential equation of the second order with an operator coefficient in Banach space

4.2.1 Introduction

The problem which we deal with in the present section is a particular case of the following initial value problem for a differential equation with operator coefficients in a Banach space:

$$\frac{d^2u(t)}{dt^2} + A\frac{du(t)}{dt} + Bu(t) = 0 \quad \text{for } t > 0, \quad \alpha > 0$$

$$u(0) = \varphi, \tag{4.36}$$

$$\frac{du(0)}{dt} = \psi,$$

which has been studied by many authors. The sufficient solvability conditions in the case of a dominant operator A were given in [38]. The uniqueness and existence theorems were proved in [61] under the so-called conditions of Miyadera–Feller–Phillips type. A classification as well as the existence and uniqueness results for solutions of problems (4.36) depending on the type of operators A and B were given in [49] (see also the literature cited therein).

A particular case of the problem (4.36) is

$$\frac{d^2u(t)}{dt^2} + \alpha A\frac{du(t)}{dt} + Au = 0, \quad \text{for } t > 0,$$

$$u(0) = \varphi, \tag{4.37}$$

$$\frac{du(0)}{dt} = \psi,$$

where A is an operator in a Banach space X equipped with norm $\|\cdot\|$. We assume that the operator $A : X \to X$ is such that its spectrum $\Sigma(A)$ lies in a segment $[\gamma_0, \infty)$ of the positive real axis, and that, for any $\epsilon > 0$, we have the resolvent estimate

$$\|(zI - A)^{-1}\| \leq C(1 + |z|)^{-1} \quad \forall z \notin \Sigma_\epsilon = \{z : |\arg(z)| < \epsilon\}, \tag{4.38}$$

with a constant $C = C_\epsilon$. In the case $A = -\Delta$, where Δ denotes the Laplacian in $\mathbb{R}^k$, $k \geq 2$ with the Dirichlet boundary condition $u(x,t) = 0$ for $x \in \partial\Omega$, and Ω being a bounded domain in $\mathbb{R}^k$, we have the so-called strongly damped wave equation [41, 66].

The finite-element approximations with a polynomial convergence for this equation were considered in [41, 66]. In the first of these works, spatial discretizations by finite elements and associated fully discrete methods were analyzed in an L_2-based norm. The analysis depends on the fact that the solution may be expressed in terms of the operator exponential. In the second work, this approach was combined with recent results on discretization to parabolic problems and essentially optimal order error bounds in maximum-norm for piecewise linear finite elements combined with the backward Euler and Crank–Nicolson time stepping methods were derived. The stability of various three-level finite-difference approximations of second-order differential equations with operator coefficients was studied in [56]. It was shown that an essential stability condition is the strong P-positiveness of the operator coefficients.

The aim of this section is to construct and to justify exponentially convergent approximations in time for problem (4.36).

4.2.2 Representation of the solution through the operator exponential

We introduce the vector w and the operator-valued matrix B by

$$w = \begin{pmatrix} u \\ u_t \end{pmatrix}, \qquad B = \begin{pmatrix} 0 & -I \\ A & \alpha A \end{pmatrix}.$$

We equip the space of vectors $v = (v_1, v_2)^T$ with the norm $|||v||| = \max\{\|v_1\|, \|v_2\|\}$, which generates the corresponding norm for operator block matrices. The problem (4.37) can be written in the form

$$\frac{dw(t)}{dt} + Bw = 0, \tag{4.39}$$
$$w(0) = w_0,$$

where

$$w_0 = \begin{pmatrix} \varphi \\ \psi \end{pmatrix}.$$

The resolvent of B is of the form

$$(zI - B)^{-1} = (\alpha z - 1)^{-1} \mathcal{R}(z) \begin{pmatrix} zI - \alpha A & -I \\ A & zI \end{pmatrix}$$
$$= (\alpha z - 1)^{-1} \begin{pmatrix} \alpha I - z(\alpha z - 1)^{-1}\mathcal{R}(z) & -\mathcal{R}(z) \\ -I + z^2(\alpha z - 1)^{-1}\mathcal{R}(z) & z\mathcal{R}(z) \end{pmatrix}, \tag{4.40}$$

where

$$\mathcal{R}(z) = (z^2(\alpha z - 1)^{-1}I - A)^{-1}.$$

Let A satisfy (4.38), then there exists $\epsilon > 0$ such that [66]

$$\|\mathcal{R}(z)\| \le \frac{C}{1 + |z|} \qquad \forall z \notin \Sigma_\epsilon = \{z : |\arg(z)| < \epsilon|\},$$

with a positive constant $C = C_\epsilon$. Using this estimate, one can show analogously to [66] that there exists an angle $\theta \in (0, \frac{\pi}{2})$ such that, on its rays and outside of the sector, the following estimate for the resolvent holds:

$$|||(zI - B)^{-1}||| \le \frac{C}{1 + |z|}, \qquad z \notin \Sigma_\theta,$$

and the angle θ contains the set

$$\Psi = \left\{z : |z - \alpha| = \alpha^{-1}, \ \mathrm{Re}z \ge \frac{1}{2}\alpha\gamma_0\right\} \bigcup \{z : \mathrm{Im}z = 0, \ \mathrm{Re}z \ge \alpha^{-1}\}.$$

For our algorithm below we need the value of θ expressed through the input data of the problem γ_0, α.

One can see from (4.40) that the spectral set $\Sigma(B)$ of the operator B consists of $z = \alpha^{-1}$ and of all z such that

$$\frac{z^2}{\alpha z - 1} - \lambda = 0,$$

where $\lambda \in \Sigma(A)$. Rewriting the last equation in the form $z^2 - \alpha\lambda z + \lambda = 0$ we obtain the function

$$z = z(\lambda) = \begin{cases} \frac{\lambda\alpha}{2} \pm \sqrt{\left(\frac{\lambda\alpha}{2}\right)^2 - \lambda}, & \text{if } \lambda \geq 4\alpha^{-2}, \\ \frac{\lambda\alpha}{2} \pm i\sqrt{\lambda - \left(\frac{\lambda\alpha}{2}\right)^2}, & \text{if } \gamma_0 \leq \lambda < 4\alpha^{-2}. \end{cases}$$

Let us consider the following cases.

1. If $\gamma_0 \geq 4\alpha^{-2}$ then we have $z = z_1(\lambda) = \frac{\lambda\alpha}{2} + \sqrt{\left(\frac{\lambda\alpha}{2}\right)^2 - \lambda}$, $z = z_2(\lambda) = \frac{\lambda\alpha}{2} - \sqrt{\left(\frac{\lambda\alpha}{2}\right)^2 - \lambda}$. The function $z_1(\lambda)$ maps the spectral set $\Sigma(A)$ into a subset of $S_1 = [2\alpha^{-1}, \infty)$. Since

$$\lim_{\lambda \to \infty} z_2(\lambda) = \lim_{\lambda \to \infty} \frac{\lambda}{\frac{\lambda\alpha}{2} + \sqrt{\left(\frac{\lambda\alpha}{2}\right)^2 - \lambda}} = \alpha^{-1}$$

and $z_2(4\alpha^{-2}) = 2\alpha^{-1}$, the function $z_2(\lambda)$ translates the set $\Sigma(A)$ into a subset of $S_2 = [\alpha^{-1}, 2\alpha^{-1}]$. Thus, the function $z(\lambda)$ maps in this case the spectral set $\Sigma(A)$ into $S_1 \cup S_2 = [\alpha^{-1}, \infty)$ provided that $\lambda \geq \gamma_0 \geq 4\alpha^{-2}$.

2. If $0 < \gamma_0 < 4\alpha^{-2}$ then we have $z = z(\lambda) = \frac{\lambda\alpha}{2} \pm i\sqrt{\lambda - \left(\frac{\lambda\alpha}{2}\right)^2}$. For such z it holds that

$$\left| z - \alpha^{-1} \right| = \left| \frac{\lambda\alpha}{2} \pm i\sqrt{\lambda - \left(\frac{\lambda\alpha}{2}\right)^2} - \alpha^{-1} \right|$$

$$= \left[\left(\frac{\lambda\alpha}{2} - \alpha^{-1} \right)^2 + \lambda - \left(\frac{\lambda\alpha}{2} \right)^2 \right]^{1/2} = \alpha^{-1},$$

i.e., these z lie in the intersection of the circle $S_3 = \{z : |z - \alpha^{-1}| = \alpha^{-1}\}$ with the half-plane $S_4 = \{z : \text{Re}\,z \geq \frac{\gamma_0\alpha}{2}\}$. Since $\frac{\alpha\gamma_0}{2} < 2\alpha^{-1}$, the point $\frac{\alpha\gamma_0}{2}$ lies inside of S_3 (see Fig. 4.1). The points A and B have the coordinates $A = (\alpha\gamma_0/2, 0)$, $B = (\alpha\gamma_0/2, \sqrt{\gamma_0 - (\gamma_0\alpha/2)^2})$ and it is easy to find that the spectral angle θ of the operator B is defined by

$$\theta = \arccos \frac{\alpha\sqrt{\gamma_0}}{2}.$$

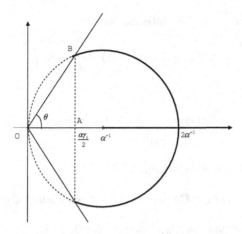

Figure 4.1: The spectral angle of the operator B.

This formula gives the upper bound for the spectral angle in all cases.

The above spectral properties of B guarantee the existence of the operator exponential e^{-Bt} and the solution of problem (4.39) can be represented by

$$w(t) = e^{-Bt}w_0.$$

4.2.3 Representation and approximation of the operator exponential

We use a convenient representation of the operator exponential as in chapter 2 by the improper Dunford-Cauchy integral

$$e^{-Bt}w_0 = \frac{1}{2\pi i} \int_{\Gamma_\Sigma} e^{-tz}(zI - B)^{-1}w_0 dz, \qquad (4.41)$$

where Γ_Σ is a path which envelopes the spectrum of the operator B. It is obvious that the spectrum of the operator B lies inside of the hyperbola

$$\Gamma_\Sigma = \{z = \cosh s - 1 - i\tan\theta \sinh s : s \in (-\infty, \infty)\}.$$

We choose a path

$$\Gamma_I = \{z = a\cosh s - 1 - ib\sinh s : s \in (-\infty, \infty)\}$$

so that it envelopes Γ_Σ and construct an algorithm like in section 3.1 (see also [27]). The particular values of a, b we define later.

For this path, we have

$$w(t) = \frac{1}{2\pi i} \int_{\Gamma_\Sigma} e^{-tz}(zI - B)^{-1}w_0 dz = \frac{1}{2\pi i} \int_{\Gamma_I} e^{-tz}(zI - B)^{-1}w_0 dz. \qquad (4.42)$$

After parametrization of the last integral, we obtain

$$w(t) = \frac{1}{2\pi i} \int_{-\infty}^{\infty} e^{-tz(s)} z'(s)(z(s)I - B)^{-1} w_0 \, ds = \int_{-\infty}^{\infty} f(t, s) \, ds, \qquad (4.43)$$

where

$$z(s) = a \cosh s - 1 - ib \sinh s,$$
$$z'(s) = a \sinh s - ib \cosh s.$$

In the following, we use the infinite strip

$$D_d := \{ z \in \mathbb{C} : -\infty < \Re e\, z < \infty, \ |\Im m\, z| < d \}$$

as well as the finite rectangles $D_d(\epsilon)$ defined for $0 < \epsilon < 1$ by

$$D_d(\epsilon) = \{ z \in \mathbb{C} : |\Re e\, z| < 1/\epsilon, \ |\Im m\, z| < d(1 - \epsilon) \}.$$

For $1 \leq p \leq \infty$, we introduce the space $\mathbf{H}^p(D_d)$ of all operator-valued functions which are analytic in D_d, such that, for each $\mathcal{F} \in \mathbf{H}^p(D_d)$, there holds $\|\mathcal{F}\|_{\mathbf{H}^p(D_d)} < \infty$ with

$$\|\mathcal{F}\|_{\mathbf{H}^p(D_d)} := \begin{cases} \lim_{\epsilon \to 0} \left(\int_{\partial D_d(\epsilon)} \|\mathcal{F}(z)\|^p |dz| \right)^{1/p} & \text{if } 1 \leq p < \infty, \\ \lim_{\epsilon \to 0} \sup_{z \in D_d(\epsilon)} \|\mathcal{F}(z)\| & \text{if } p = \infty. \end{cases}$$

For our further analysis and for the algorithm below it is necessary, analogously to section 3.1, to find the width of a strip D_d to which we can analytically extend the function $f(t, s)$ with respect to s. The change of variable s to $s + i\nu$ maps the integration hyperbola Γ_I into the parametric set

$$\Gamma(\nu) = \{ z = a \cosh(s + i\nu) - 1 - ib \sinh(s + i\nu) : s \in (-\infty, \infty) \}$$
$$= \{ z = (a \cos \nu + b \sin \nu) \cosh s - 1 - i(b \cos \nu - a \sin \nu) \sinh s : s \in (-\infty, \infty) \}.$$

We have to choose the parameters a, b so that for $\nu \in \left(-\frac{d}{2}, \frac{d}{2} \right)$, $\Gamma(\nu)$ does not intersect Γ_Σ and, at the same time, envelopes the spectrum. Let us select a, b in the following way: when $\nu = -\frac{d}{2}$, $\Gamma(\nu)$ should be transformed into a line parallel to the imaginary axis, and for $\nu = \frac{d}{2}$ the set $\Gamma(\nu)$ should coincide with Γ_Σ. These requirements provide the equations

$$\begin{cases} a \cos \frac{d}{2} - b \sin \frac{d}{2} = 0, \\ a \cos \frac{d}{2} + b \sin \frac{d}{2} = 1, \\ b \cos \frac{d}{2} - a \sin \frac{d}{2} = \tan \theta. \end{cases}$$

The solution of this system is as follows:

$$\begin{cases} d = \dfrac{\pi}{2} - \theta, \\[2mm] a = \dfrac{1}{2\cos\left(\frac{\pi}{4} - \frac{\theta}{2}\right)}, \\[2mm] b = \dfrac{1}{2\sin\left(\frac{\pi}{4} - \frac{\theta}{2}\right)}. \end{cases} \tag{4.44}$$

For these parameters, the vector-valued function $f(t, s)$ is analytic with respect to s in the strip

$$D_d = \left\{ \zeta = \xi + i\nu \; : \; \xi \in (-\infty, \infty), \; |\nu| < \tfrac{d}{2} \right\}.$$

To provide the numerical stability of the algorithm, we modify the resolvent in (4.42) analogously to section 3.1. The only difference from section 3.1 is that we add $\frac{1}{z+1}I$ instead of $\frac{1}{z}I$ as in section 3.1 since our integration hyperbola passes through the origin. Thus, we represent the solution of (4.39) as

$$\begin{aligned} w(t) &= \frac{1}{2\pi i} \int_{\Gamma_I} e^{-tz} \left[(zI - B)^{-1} - \frac{1}{z+1}I \right] w_0 \, dz \\ &= \int_{-\infty}^{\infty} f_1(t, s)\, ds, \end{aligned} \tag{4.45}$$

where

$$f_1(t, s) = \frac{1}{2\pi i} e^{-tz(s)} z'(s) \left[(z(s)I - B)^{-1} - \frac{1}{z(s)+1} \right] w_0.$$

Remark 4.4. Note that

$$\int_{\Gamma_I} e^{-tz} \frac{1}{z+1}\, ds = 0,$$

because the point $(z+1)$ remains outside of the integration path, i.e., our correction does not change the value of the integral, but influences the stability of the algorithm below (see section 3.1 for details).

Let us estimate the function $f_1(t, s)$ in (4.45) under the additional assumption that $w_0 \in D(B^\sigma)$ for some $\sigma > 0$. It is easy to see that this means that φ, $\psi \in D(A^\sigma)$. Analogously to section 3.1 we obtain

$$\begin{aligned} \left\| \left[(z(s)I - B)^{-1} - \frac{1}{z(s)+1}I \right] w_0 \right\| &= \frac{1}{|z(s)+1|} \left\| (B+I)(z(s)I - B)^{-1} w_0 \right\| \\ &\leq \frac{1}{|z(s)+1|} \left\| (B+I)B^{-1} \right\| \left\| B(z(s)I - B)^{-1} w_0 \right\| \\ &\leq \frac{C}{|z(s)+1|} \left\| B^{1-\sigma}(z(s)I - B)^{-1} \right\| \left\| B^\sigma w_0 \right\| \\ &\leq \frac{C}{|z(s)+1|\,(1+|z(s)|)^\sigma} \left\| B^\sigma w_0 \right\|. \end{aligned}$$

with positive constants C. Taking into account this inequality we arrive at the following estimate of the integrand:

$$
\begin{aligned}
\|f_1(t,s)\| &\leq \frac{C|\mathrm{e}^{-z(s)t}|\,|z'(s)|}{2\pi|z(s)+b|\,(1+|z(s)|)^\sigma}\,\|B^\sigma w_0\| \\[2mm]
&\leq C_1 \mathrm{e}^{-ta\cosh s+t}\,\frac{\sqrt{a^2\sinh^2 s + b^2\cosh^2 s}}{\left(a^2\cosh^2 s + b^2\sinh^2 s\right)^{\frac{\sigma+1}{2}}}\,\|B^\sigma w_0\| \\[2mm]
&\leq C_1 \mathrm{e}^{-ta\cosh s+t}\left(\frac{a^2\tanh^2 s + b^2}{a^2+b^2\tanh^2 s}\right)^{\frac{1}{2}}\frac{\|B^\sigma w_0\|}{(\cosh s)^\sigma\left(a^2+b^2\tanh^2 s\right)^{\frac{\sigma}{2}}} \\[2mm]
&\leq C_2 \mathrm{e}^{-ta\cosh s+t-\sigma|s|}\,\|B^\sigma w_0\| \qquad\qquad\qquad\qquad (4.46)
\end{aligned}
$$

with a new positive constant C_1. Estimate (4.46) shows that integral (4.45) exists (is convergent) $\forall\, t \geq 0$, provided that $\sigma > 0$.

Proceeding analogously to section 3.1, we obtain

$$
\|f_1(t,\cdot)\|_{H^1(D_d)} \leq C(\theta,\sigma)\int_{-\infty}^{\infty} \mathrm{e}^{-\sigma|s|}ds = \frac{2C(\theta,\sigma)}{\sigma},
$$

where

$$
C(\theta,\sigma) = C_2\left(\frac{2}{a}\right)^\sigma\|B^\sigma w_0\| = C_2\frac{1}{\left[\cos\left(\frac{\pi}{4}-\frac{\theta}{2}\right)\right]^\sigma}\|B^\sigma w_0\|.
$$

To approximate the integral (4.45) let us use the following Sinc quadrature [62]:

$$
w(t) \approx w_N(t) = \frac{h}{2\pi\mathrm{i}}\sum_{k=-N}^{N} f_1(t, z(kh)). \qquad\qquad (4.47)
$$

For the error of this quadrature, we have

$$
\begin{aligned}
\|\eta_N(f_1, h)\| &= \|w(t) - w_N(t)\| \\[2mm]
&\leq \left\|w(t) - \frac{h}{2\pi\mathrm{i}}\sum_{k=-N}^{N} f_1(t, z(kh))\right\| + \left\|\frac{h}{2\pi\mathrm{i}}\sum_{|k|>N} f_1(t, z(kh))\right\|.
\end{aligned}
$$

For the first sum in this inequality the following estimate holds true (see [23, 30, 18] for details):

$$
\begin{aligned}
\left\|w(t) - \frac{h}{2\pi\mathrm{i}}\sum_{k=-N}^{N} f_1(t, z(kh))\right\| &\leq \frac{1}{2\pi}\frac{\mathrm{e}^{-\pi d/h}}{2\sinh \pi d/h}\|f_1(t,v)\|_{H^1(D_d)} \\[2mm]
&\leq \frac{C(\theta,\sigma)}{2\pi\sigma}\frac{\mathrm{e}^{-\pi d/h}}{2\sinh \pi d/h}. \qquad (4.48)
\end{aligned}
$$

For the second summand, we have

$$
\left\| \frac{h}{2\pi i} \sum_{|k|>N} f_1(t, z(kh)) \right\| \leq \frac{h}{2\pi} \sum_{|k|>N} \| f_1(t, z(kh)) \|
$$

$$
\leq \frac{h}{\pi} \sum_{k=N+1}^{\infty} C(\theta, \sigma) e^{-ta\cosh(kh)+t-kh\sigma}
$$

$$
\leq \begin{cases} \dfrac{C(\theta, \sigma) e^t}{\pi t} e^{-ta\cosh((N+1)h)-\sigma(N+1)h}, & t > 0, \\[2mm] C(\theta, \sigma) e^{-\sigma(N+1)h}, & t = 0. \end{cases} \tag{4.49}
$$

Combining the estimates (4.48) and (4.49) for $t = 0$, we conclude that

$$
\| \eta_N(f_1, h) \| \leq \frac{c}{\sigma} \left[\frac{e^{-\pi d/h}}{\sinh \pi d/h} + e^{-\sigma(N+1)h} \right]. \tag{4.50}
$$

Equalizing both exponentials by the choice

$$
h = \sqrt{\frac{\pi d}{\sigma(N+1)}}, \tag{4.51}
$$

we obtain

$$
\| \eta_N(f_1, h) \| \leq \frac{c}{\sigma} \left[\frac{e^{-\sqrt{\pi d\sigma(N+1)}}}{\sinh\left(\sqrt{\pi d\sigma(N+1)}\right)} + e^{-\sqrt{\pi d\sigma(N+1)}} \right]
$$

$$
= \frac{c}{\sigma} \left[\frac{e^{-2\sqrt{\pi d\sigma(N+1)}}}{1 - e^{-\sqrt{\pi d\sigma(N+1)}}} + e^{-\sqrt{\pi d\sigma(N+1)}} \right]. \tag{4.52}
$$

In the case $t > 0$, the first summand in the exponent of $\exp(-ta\cosh((N+1)h) - \sigma|(N+1)h|)$ in (4.49) contributes mainly to the error order. Setting $h = O(\ln N/N)$, we remain for a fixed $t > 0$ asymptotically with an error

$$
\| \eta_N(f_1, h) \| \leq c \left[\frac{e^{-c_1 N/\ln N}}{\sigma} + \frac{e^{-tc_2 N/\ln N}}{t} \right], \tag{4.53}
$$

where c, c_1, c_2 are positive constants.

Thus, we have proven the following result.

Theorem 4.5. *Let $\varphi, \psi \in D(A^\sigma)$ and A be an operator satisfying estimate (4.38) and having spectrum $\Sigma(A)$ which lies in a segment $[\gamma_0, \infty)$ of the positive real axis. Then the formula (4.47) represents an approximate solution of the initial value problem (4.39) with estimate (4.52) of the order $\mathcal{O}(e^{-c\sqrt{N}})$ provided that $t \geq 0$, $h = \sqrt{\frac{\pi d}{\sigma(N+1)}}$ and with the estimate (4.53) of the order $\mathcal{O}(e^{-cN/\ln N})$ provided that $t > 0$, $h = \mathcal{O}(\ln N/N)$.*

4.2.4 Numerical examples

In this section, we consider a model problem of type (4.39) with a known explicit solution, solve this problem numerically by our algorithm and compare with theoretical results predicted.

Let us consider the following initial boundary value problem:

$$
\begin{aligned}
&\frac{\partial^2 u(t,x,y)}{\partial t^2} - \alpha \frac{\partial}{\partial t}\left(\Delta u(t,x,y)\right) - \Delta u(t,x,y) = 0, \\
&u(t,0,y) = u(t,1,y) = u(t,x,0) = u(t,x,1) = 0, \\
&u(0,x,y) = \sin(\pi x)\sin(2\pi y), \\
&\left.\frac{\partial u(t,x,y)}{\partial t}\right|_{t=0} = 0,
\end{aligned}
\tag{4.54}
$$

where

$$
\Delta u(t,x,y) = \frac{\partial^2 u(t,x,y)}{\partial x^2} + \frac{\partial^2 u(t,x,y)}{\partial y^2}.
$$

Depending on parameter α, there are three cases of the solution of problem (4.54).

1. If $0 < \alpha < \frac{2}{\pi\sqrt{5}}$, then

$$
u(t,x,y,\alpha) = \exp\{p_1 t\}\left(\cos(p_2 t) - \frac{p_1}{p_2}\sin(p_2 t)\right)\sin(\pi x)\sin(2\pi y),
$$

where

$$
p_1 = -\frac{5}{2}\pi^2\alpha,
$$

$$
p_2 = \frac{1}{2}\sqrt{20\pi^2 - 25\pi^4\alpha^2}.
$$

2. If $\alpha = \frac{2}{\pi\sqrt{5}}$, then

$$
u(t,x,y,\alpha) = \exp\left\{-\sqrt{5}\pi t\right\}\left(1 + \pi\sqrt{5}t\right)\sin(\pi x)\sin(2\pi y).
$$

3. If $\alpha > \frac{2}{\pi\sqrt{5}}$, then

$$
u(t,x,y,\alpha) = \left(\exp\{p_1 t\}p_2 - \exp\{p_2 t\}p_1\right)\frac{\sin(\pi x)\sin(2\pi y)}{p_2 - p_1},
$$

where

$$
p_1 = -\frac{5}{2}\pi^2\alpha + \frac{1}{2}\sqrt{25\pi^4\alpha^2 - 20\pi^2},
$$

$$
p_2 = -\frac{5}{2}\pi^2\alpha - \frac{1}{2}\sqrt{25\pi^4\alpha^2 - 20\pi^2}.
$$

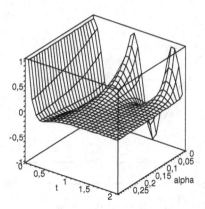

Figure 4.2: $\alpha < \frac{2}{\pi\sqrt{5}}$.

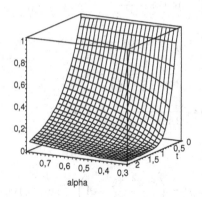

Figure 4.3: $\alpha > \frac{2}{\pi\sqrt{5}}$.

The solutions of problem (4.54) are presented in Figures 4.2–4.4 depending on parameter α, at the point $(x, y) = (\frac{1}{2}, \frac{1}{4})$.

Example 4.6. In the case $\alpha = \frac{2}{\pi\sqrt{5}}$, it is easy to check that $\gamma_0 = 2\pi^2$ and

$$\alpha = \frac{2}{\pi\sqrt{5}} < \frac{2}{\sqrt{\gamma_0}} = \frac{\sqrt{2}}{\pi}.$$

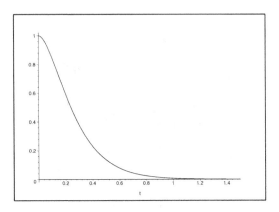

Figure 4.4: $\alpha = \frac{2}{\pi\sqrt{5}}$.

Due to (4.44) and definition of θ, we obtain

$$\theta = \arccos \frac{\frac{2}{\pi\sqrt{5}}\sqrt{2\pi^2}}{2} = \arccos\sqrt{\frac{2}{5}},$$

$$d = \frac{\pi}{2} - \arccos\frac{\sqrt{2}}{5} \approx 0,68,$$

$$a = \frac{1}{2\cos\left(\frac{\pi}{4} - \frac{\theta}{2}\right)} \approx 0,53,$$

$$b = \frac{1}{2\sin\left(\frac{\pi}{4} - \frac{\theta}{2}\right)} \approx 1,49.$$

Using

$$\mathcal{R}(z)\sin(\pi x)\sin(2\pi y) = \left(\frac{z^2}{\alpha z - 1}I - A\right)^{-1}\sin(\pi x)\sin(2\pi y)$$

$$= \frac{1}{\frac{z^2}{\alpha z - 1} - 5\pi^2}\sin(\pi x)\sin(2\pi y) = \frac{\alpha z - 1}{z^2 - 5\pi^2\alpha z + 5\pi^2}\sin(\pi x)\sin(2\pi y),$$

we can obtain explicitly

$$(zI - B)^{-1}w_0 = \begin{pmatrix} \left(\alpha - \dfrac{z}{z^2 - 5\pi^2\alpha z + 5\pi^2}\right)\dfrac{\sin(\pi x)\sin(2\pi y)}{\alpha z - 1} \\[3ex] \left(-1 + \dfrac{z^2}{z^2 - 5\pi^2\alpha z + 5\pi^2}\right)\dfrac{\sin(\pi x)\sin(2\pi y)}{\alpha z - 1} \end{pmatrix}$$

and further apply algorithm (4.47). The computations were provided with Maple 9. The results of computation presented in Table 4.1 confirm our theory above.

N	ε
4	0.948769502e-1
8	0.300504394e-1
16	0.5866423681025070l8e-2
32	0.53379300563269739e-3
64	0.17784977714518990e-4
128	0.14181989945806e-6
256	0.14977670055e-9

Table 4.1: The error in the case $\alpha = \frac{2}{\pi\sqrt{5}}$, $t = 0.5$, $x = 0.5$, $y = 0.2$.

Example 4.7. In the second example we consider the problem (4.54) with $\alpha = \frac{2}{5\pi}$. The exact solution in this case is

$$u(t, x, y) = \left(\cos(2\pi t) + \frac{1}{2}\sin(2\pi t)\right) e^{-\pi t} \sin(\pi x) \sin(2\pi y).$$

Exactly as in Example 4.6 we have $\gamma_0 = 2\pi^2$,

$$\alpha = \frac{2}{5\pi} < \frac{2}{\sqrt{\gamma_0}} = \frac{\sqrt{2}}{\pi}.$$

Due to (4.44) and to the definition of θ, we obtain

$$\theta = \arccos \frac{\frac{2}{5\pi}\sqrt{2\pi^2}}{2} = \arccos \frac{\sqrt{2}}{5},$$

$$d = \frac{\pi}{2} - \arccos \frac{\sqrt{2}}{5} \approx 0,29,$$

$$a = \frac{1}{2\cos\left(\frac{\pi}{4} - \frac{\theta}{2}\right)} \approx 0,51,$$

$$b = \frac{1}{2\sin\left(\frac{\pi}{4} - \frac{\theta}{2}\right)} \approx 3,5.$$

The function $\mathcal{R}(z)$ and the resolvent $(zI - B)^{-1} w_0$ for this case can be calculated explicitly analogously to Example 4.6. Further computations in accordance with (4.47) were provided with Maple 9. The results of computation are presented in Table 4.2 and confirm our theory above.

4.3 Exponentially convergent approximation to the elliptic solution operator

4.3.1 Elliptic problems in cylinder type domains

We consider an elliptic problem in a cylindrical domain.

N	ε
8	0.1557679918
16	0.5180574507494855008e-1
32	0.1097312605636203481e-1
64	0.121349430767272743e-2
128	0.5330597600436985e-4
256	0.63049289198752e-6

Table 4.2: The error in the case $\alpha = \frac{2}{5\pi}$, $t = 0.5$, $x = 0.5$, $y = 0.2$.

Let A be a linear, densely defined, closed, strongly positive operator in a Banach space X. The operator-valued function (hyperbolic sine family of bounded operators [21])

$$E(x) \equiv E(x; \sqrt{A}) := \sinh^{-1}(\sqrt{A}) \sinh(x\sqrt{A}),$$

satisfies the elliptic differential equation

$$\frac{d^2 E}{dx^2} - AE = 0, \quad E(0) = \Theta, \quad E(1) = I,$$

where I is the identity and Θ the zero operator. Given the normalized hyperbolic operator sine family $E(x)$, the solution of the homogeneous elliptic differential equation (elliptic equation)

$$\frac{d^2 u}{dx^2} - Au = 0, \quad u(0) = 0, \quad u(1) = u_1 \tag{4.55}$$

with a given vector u_1 and the unknown vector-valued function $u(x) : (0,1) \to X$ can be represented as

$$u(x) = E(x; \sqrt{A})u_1.$$

In applications, A can be an elliptic partial differential operator with respect to the spatial variables $(x_1, x_2, \ldots, x_d) \in \Omega_A \subset \mathbb{R}^d$ (see, e.g., Example 1). The solution of the inhomogeneous elliptic boundary value problem posed in the cylinder type domain $\Omega_A \times [0,1] \subset \mathbb{R}^{d+1}$,

$$\frac{d^2 u}{dx^2} - Au = -f(x), \quad u(0) = u_0, \quad u(1) = u_1, \tag{4.56}$$

with the unknown function $u(x) = u(x; x_1, \ldots, x_d)$, with known boundary data u_0, u_1 and with the given right-hand side $f(x) = f(x; x_1, \ldots, x_d)$, can be represented as

$$u(x) = u_h(x) + u_p(x), \tag{4.57}$$

where

$$u_h(x) = E(x; \sqrt{A})u_1 + E(1 - x; \sqrt{A})u_0 \tag{4.58}$$

is the solution of the homogeneous problem and

$$u_p(x) = \int_0^1 G(x, s; A) f(s) ds,$$ (4.59)

is the solution of the inhomogeneous problem with the Green function

$$G(x, s; A) \equiv G(x, s)$$

$$= [\sqrt{A} \sinh \sqrt{A}]^{-1} \begin{cases} \sinh(x\sqrt{A}) \sinh((1-s)\sqrt{A}) & x \leq s, \\ \sinh(s\sqrt{A}) \sinh((1-x)\sqrt{A}) & x \geq s. \end{cases}$$ (4.60)

4.3.2 New algorithm for the normalized operator sinh-family

We consider the following representation of the solution of problem (4.55)

$$u(x) = u_{hl}(x) = \frac{1}{2\pi i} \int_{\Gamma_I} E(x; \sqrt{z})(zI - A)^{-1} u_1 dz,$$ (4.61)

where

$$E(x; \sqrt{z}) = \sinh(x\sqrt{z}) / \sinh \sqrt{z}$$

is the normalized hyperbolic sine function. Our goal consists of an approximation of this integral by a quadrature with the exponential convergence rate including $x = 1$. It is of great importance to have in mind the representation of the solution of the non-homogeneous boundary value problem (4.56)–(4.60), where the argument of the hyperbolic operator sine family under the integral becomes zero for $x = s = 0$ or $x = s = 1$. Taking into account (2.27) for $m = 0$, we can represent

$$u(x) = u_{hl}(x) = \frac{1}{2\pi i} \int_{\Gamma_I} E(x; \sqrt{z}) \left[(zI - A)^{-1} - \frac{1}{z} I \right] u_1 dz,$$ (4.62)

instead of (4.61) (for $x > 0$ the integral from the second summand is equal to zero due to the analyticity of the integrand inside of the integration path) and this integral represents the solution of the problem (4.55) for $u_1 \in D(A^\alpha)$, $\alpha > 0$.

Similar to that in section 3.1.2, we call the hyperbola

$$\Gamma_0 = \{z(\xi) = a_0 \cosh \xi - ib_0 \sinh \xi : \xi \in (-\infty, \infty), \ b_0 = a_0 \tan \varphi\}$$

the spectral hyperbola, which goes through the vertex $(a_0, 0)$ of the spectral angle and possesses asymptotes which are parallel to the rays of the spectral angle Σ (see Fig. 3.1). We choose a hyperbola

$$\Gamma_I = \{z(\xi) = a_I \cosh \xi - ib_I \sinh \xi : \xi \in (-\infty, \infty)\}$$ (4.63)

as an integration path, where the parameter will be chosen later:

$$u(x) = u_{hl}(x) = \frac{1}{2\pi i} \int_{-\infty}^{\infty} \mathcal{F}(x, \xi) d\xi,$$ (4.64)

with

$$\mathcal{F}(x,\xi) = F_A(x,\xi)u_1,$$

$$F_A(x,\xi) = E(x;\sqrt{z(\xi)})(a_I \sinh \xi - ib_I \cosh \xi)\left[(z(\xi)I - A)^{-1} - \frac{1}{z(\xi)}I\right].$$

In order to estimate $\|\mathcal{F}(x,\xi)\|$ we need estimates for

$$|z'(\xi)/z(\xi)| = (a_I \sinh \xi - ib_I \cosh \xi)/(a_I \cosh \xi - ib_I \sinh \xi)$$

$$= \sqrt{(a_I^2 \tanh^2 \xi + b_I^2)/(b_I^2 \tanh^2 \xi + a_I^2)}$$

and for $\|E(x;\sqrt{z(\xi)})\|$.

It was shown in section 3.1 that

$$|z'(\xi)/z(\xi)| \le b_I/a_I. \tag{4.65}$$

Before we move to estimating $\|E(x;\sqrt{z(\xi)})\|$, let us estimate $|\sqrt{z(\xi)}|$. We have

$$|\sqrt{z(\xi)}| = |\sqrt{a_I \cosh \xi - ib_I \sinh \xi}| = \sqrt[4]{a_I^2 \cosh^2 \xi + b_I^2 \sinh^2 \xi}$$

$$= \sqrt[2]{a_I \cosh \xi}\sqrt[4]{1 + (b_I/a_I)^2 \tanh^2 \xi},$$

from which we obtain

$$\sqrt{a_I \cosh \xi} \le |\sqrt{z(\xi)}| \le \sqrt[4]{1 + (b_I/a_I)^2}\sqrt{a_I \cosh \xi}. \tag{4.66}$$

Using this estimate we have

$$|e^{-2x\sqrt{z}}| = \left|\exp\left\{-2x\left(\sqrt{\left(a_I \cosh \xi + \sqrt{a_I^2 \cosh^2 \xi + b_I^2 \sinh^2 \xi}\right)/2}\right.\right.\right.$$

$$\left.\left.\left. +i\sqrt{\left(-a_I \cosh \xi + \sqrt{a_I^2 \cosh^2 \xi + b_I^2 \sinh^2 \xi}\right)/2}\right)\right\}\right|$$

$$= \exp\left\{-2x\left(\sqrt{\left(a_I \cosh \xi + \sqrt{a_I^2 \cosh^2 \xi + b_I^2 \sinh^2 \xi}\right)/2}\right)\right\}$$

$$< e^{-2x\sqrt{a_I \cosh \xi}} \quad \forall x \in (0,1], \tag{4.67}$$

$$|1 - e^{-2x\sqrt{z}}| \le 1 + |e^{-2x\sqrt{z}}| \le 2,$$

$$|1 - e^{-2x\sqrt{z}}| \ge 1 - |e^{-2x\sqrt{z}}| \ge 1 - e^{-2x\sqrt{a_I \cosh \xi}},$$

$$|e^{(x-1)\sqrt{z}} - e^{-(x+1)\sqrt{z}}| = |e^{(x-1)\sqrt{z}}(1 - e^{-2x\sqrt{z}})| \le 2e^{(x-1)\sqrt{a_I \cosh \xi}},$$

and, finally,

$$|E(x; \sqrt{z})| = \left| \frac{e^{x\sqrt{z}} - e^{-x\sqrt{z}}}{e^{\sqrt{z}} - e^{-\sqrt{z}}} \right| = \left| \frac{e^{(x-1)\sqrt{z}} - e^{-(x+1)\sqrt{z}}}{1 - e^{-2\sqrt{z}}} \right|$$

$$\leq \frac{2e^{(x-1)\sqrt{a_I} \cosh \xi}}{1 - e^{-2\sqrt{a_I} \cosh \xi}} \leq \frac{2}{1 - e^{-2\sqrt{a_I}}} e^{(x-1)\sqrt{a_I} \cosh \xi}.$$

Supposing $u_1 \in D(A^\alpha)$, $0 < \alpha < 1$, using (4.65) and Theorem 2.4, we can estimate the integrand on the real axis $\xi \in \mathbb{R}$ for each $x \in (0,1)$ by

$$\|\mathcal{F}(x, \xi)\| = \|F_A(x, \xi)u_1\| \leq |E(x; \sqrt{z})| \cdot \left| \frac{z'(\xi)}{z(\xi)} \right| \frac{(1+M)K}{|z(\xi)|^\alpha} \|A^\alpha u_1\|$$

$$\leq (1+M)K \frac{b_I}{a_I} \frac{|E(x; \sqrt{z})|}{|z(\xi)|^\alpha} \|A^\alpha u_1\|$$

$$\leq (1+M)K \frac{2b_I}{a_I(1 - e^{-2\sqrt{a_I}})} \frac{e^{(x-1)\sqrt{a_I} \cosh \xi}}{|z(\xi)|^\alpha} \|A^\alpha u_1\| \tag{4.68}$$

$$\leq (1+M)K \frac{b_I}{1 - e^{-2\sqrt{a_I}}} \left(\frac{2}{a_I} \right)^{1+\alpha} e^{(x-1)\sqrt{a_I} \cosh \xi - \alpha|\xi|} \|A^\alpha u_1\|,$$

$$\xi \in \mathbb{R}, \quad x \in (0,1].$$

Let us show that the function $\mathcal{F}(x, \xi)$ can be analytically extended into a strip of a width d_1 with respect to ξ. After changing ξ to $\xi + i\nu$, the integration hyperbola Γ_I will be translated into the curve

$$\Gamma(\nu) = \{z(w) = a_I \cosh (\xi + i\nu) - ib_I \sinh (\xi + i\nu) : \xi \in (-\infty, \infty)\}$$
$$= \{z(w) = a(\nu) \cosh \xi - ib(\nu) \sinh \xi : \xi \in (-\infty, \infty)\}$$

with

$$a(\nu) = a_I \cos \nu + b_I \sin \nu = \sqrt{a_I^2 + b_I^2} \sin (\nu + \phi/2),$$

$$b(\nu) = b_I \cos \nu - a_I \sin \nu = \sqrt{a_I^2 + b_I^2} \cos (\nu + \phi/2), \tag{4.69}$$

$$\cos \frac{\phi}{2} = \frac{b_I}{\sqrt{a_I^2 + b_I^2}}, \quad \sin \frac{\phi}{2} = \frac{a_I}{\sqrt{a_I^2 + b_I^2}}.$$

The analyticity of the function $\mathcal{F}(x, \xi + i\nu)$, $|\nu| < d_1/2$ can be violated when the resolvent becomes unbounded. Thus, we must choose d_1 so that the hyperbola $\Gamma(\nu)$ for $\nu \in (-d_1/2, d_1/2)$ remains in the right half-plane of the complex plane, for $\nu = -d_1/2$ coincides with the imaginary axis, for $\nu = d_1/2$ coincides with the spectral hyperbola and for all $\nu \in (-d_1/2, d_1/2)$ does not intersect the spectral sector. Then we can choose the hyperbola $\Gamma(0)$ as the integration hyperbola.

This implies the system of equations

$$\begin{cases} a_I \cos\left(d_1/2\right) + b_I \sin\left(d_1/2\right) = a_0, \\ b_I \cos\left(d_1/2\right) - a_I \sin\left(d_1/2\right) = a_0 \tan\varphi, \\ a_I \cos\left(-d_1/2\right) + b_I \sin\left(-d_1/2\right) = 0, \end{cases}$$

which yields

$$\begin{cases} 2a_I \cos\left(d_1/2\right) = a_0, \\ b_I = a_0 \sin\left(d_1/2\right) + b_0 \cos\left(d_1/2\right), \\ a_I = a_0 \cos\left(d_1/2\right) - b_0 \sin\left(d_1/2\right). \end{cases}$$

Eliminating a_I from the first and the third equations we get $a_0 \cos d_1 = b_0 \sin d_1$, i.e., $d_1 = \pi/2 - \varphi$ with $\cos\varphi = \frac{a_0}{\sqrt{a_0^2+b_0^2}}$, $\sin\varphi = \frac{b_0}{\sqrt{a_0^2+b_0^2}}$ and φ being the spectral angle. Thus, if we choose the parameters of the integration hyperbola by

$$
\begin{aligned}
a_I &= a_0 \cos\left(\frac{\pi}{4} - \frac{\varphi}{2}\right) - b_0 \sin\left(\frac{\pi}{4} - \frac{\varphi}{2}\right) \\
&= \sqrt{a_0^2 + b_0^2} \cos\left(\frac{\pi}{4} + \frac{\varphi}{2}\right) = a_0 \frac{\cos\left(\frac{\pi}{4} + \frac{\varphi}{2}\right)}{\cos\varphi}, \\
b_I &= a_0 \sin\left(\frac{\pi}{4} - \frac{\varphi}{2}\right) + b_0 \cos\left(\frac{\pi}{4} - \frac{\varphi}{2}\right) \\
&= \sqrt{a_0^2 + b_0^2} \sin\left(\frac{\pi}{4} + \frac{\varphi}{2}\right) = a_0 \frac{\sin\left(\frac{\pi}{4} + \frac{\varphi}{2}\right)}{\cos\varphi},
\end{aligned}
\tag{4.70}
$$

then the vector-valued function $\mathcal{F}(x, w)$ is for all $x \in [0, 1]$ analytic with respect to $w = \xi + i\nu$ in the strip

$$D_{d_1} = \{w = \xi + i\nu : \ \xi \in (-\infty, \infty), \ |\nu| < d_1/2\}.$$

Comparing (4.70) with (4.69), we observe that $\phi = \pi/2 - \varphi$ and

$$a(\nu) = a_I \cos\nu + b_I \sin\nu = \frac{a_0 \sin\left(\nu + \pi/4 - \varphi/2\right)}{\cos\varphi} = \frac{a_0 \cos\left(\pi/4 + \varphi/2 - \nu\right)}{\cos\varphi},$$

$$b(\nu) = b_I \cos\nu - a_I \sin\nu = \frac{a_0 \sin\left(\pi/4 + \varphi/2 - \nu\right)}{\cos\varphi}.$$

We choose a positive δ such that $\varphi + \frac{\delta}{2} < \frac{\pi}{2} - \frac{\delta}{2}$, set $d = d_1 - \delta$ and consider ν such that $|\nu| < \frac{d}{2} = \frac{\pi}{4} - \frac{\varphi}{2} - \frac{\delta}{2}$. Note that $\delta \to 0$ when $\varphi \to \pi/2$. Since $\varphi + \frac{\delta}{2} \le \frac{\pi}{4} + \frac{\varphi}{2} - \nu \le \frac{\pi}{2} - \frac{\delta}{2}$, we have

$$
\begin{aligned}
\frac{a_0 \cos\left(\frac{\pi}{2} - \frac{\delta}{2}\right)}{\cos\varphi} &\le a(\nu) = \frac{a_0 \cos\left(\pi/4 + \varphi/2 - \nu\right)}{\cos\varphi} \le \frac{a_0 \cos\left(\varphi + \frac{\delta}{2}\right)}{\cos\varphi}, \\
\frac{a_0 \sin\left(\varphi + \frac{\delta}{2}\right)}{\cos\varphi} &\le b(\nu) = \frac{a_0 \sin\left(\pi/4 + \varphi/2 - \nu\right)}{\cos\varphi} \le \frac{a_0 \sin\left(\frac{\pi}{2} - \frac{\delta}{2}\right)}{\cos\varphi}.
\end{aligned}
\tag{4.71}
$$

Replacing a_I and b_I by $a(\nu)$ and $b(\nu)$ in (4.68) and taking into account (4.71), we arrive at the estimate

$$\|\mathcal{F}(x, w)\| = \|F_A(x, w)u_1\|$$

$$\leq \frac{(1 + M)Ka_0 \sin\left(\frac{\pi}{2} - \frac{\delta}{2}\right)}{\cos\varphi \left(1 - e^{-2\sqrt{a_0 \cos\left(\frac{\pi}{2} - \frac{\delta}{2}\right)/\cos\varphi}}\right)}$$

$$\times \left(\frac{2\cos\varphi}{a_0 \cos\left(\frac{\pi}{2} - \frac{\delta}{2}\right)}\right)^{1+\alpha} e^{(x-1)\tilde{a}\sqrt{\cosh\xi} - \alpha|\xi|} \|A^\alpha u_1\|,$$

$$\xi \in \mathbb{R}, \ \forall x \in [0, 1], \ \forall w \in D_d,$$

where

$$\tilde{a} = \sqrt{a_0 \cos\left(\varphi + \delta/2\right)/\cos\varphi}.$$

Taking into account that the integrals over the vertical sides of the rectangle $D_d(\epsilon) = \{z \in \mathbb{C} : |\Re e\, z| < 1/\epsilon, \ |\Im m\, z| < d(1 - \epsilon)\}$ with the boundary $\partial D_d(\epsilon)$ vanish as $\epsilon \to 0$, this estimate implies

$$\|\mathcal{F}(x, \cdot)\|_{\mathbf{H}^1(D_d)} = \lim_{\epsilon \to 0} \left(\int_{\partial D_d(\epsilon)} \|\mathcal{F}(z)\| |dz|\right)$$

$$\leq C(\varphi, \alpha, \delta)\|A^\alpha u_1\| \int_{-\infty}^{\infty} e^{-\alpha|\xi|} d\xi = \frac{2}{\alpha} C(\varphi, \alpha, \delta)\|A^\alpha u_0\| \tag{4.72}$$

with

$$C(\varphi, \alpha, \delta) = \frac{2(1 + M)Ka_0 \sin\left(\frac{\pi}{2} - \frac{\delta}{2}\right)}{\cos\varphi \left(1 - e^{-2\sqrt{a_0 \cos\left(\frac{\pi}{2} - \frac{\delta}{2}\right)/\cos\varphi}}\right)} \left(\frac{2\cos\varphi}{a_0 \cos\left(\frac{\pi}{2} - \frac{\delta}{2}\right)}\right)^{1+\alpha}.$$

Note that the constant $C(\varphi, \alpha, \delta)$ tends to ∞ if $\alpha \to 0$ or $\varphi \to \pi/2$.

We approximate integral (4.64) by the Sinc-quadrature

$$u_N(t) = u_{hl,N}(x) = \frac{h}{2\pi i} \sum_{k=-N}^{N} \mathcal{F}(t, z(kh)) \tag{4.73}$$

with the error

$$\|\eta_N(\mathcal{F}, h)\| = \|u(t) - u_N(t)\| \leq \eta_{1,N}(\mathcal{F}, h) + \eta_{2,N}(\mathcal{F}, h), \tag{4.74}$$

where

$$\eta_{1,N}(\mathcal{F}, h) = \left\|u(t) - \frac{h}{2\pi i} \sum_{k=-\infty}^{\infty} \mathcal{F}(t, z(kh))\right\|,$$

$$\eta_{2,N}(\mathcal{F}, h) = \left\|\frac{h}{2\pi i} \sum_{|k|>N} \mathcal{F}(t, z(kh))\right\|.$$

The first term can be estimated by (see Theorem 3.2.1, p. 144 in [62])

$$\|\eta_{1,N}(\mathcal{F}, h)\| \leq \frac{1}{2\pi} \frac{e^{-\pi d/h}}{2\sinh{(\pi d/h)}} \|\mathcal{F}\|_{\mathbf{H}^1(D_d)}. \tag{4.75}$$

For the second term we have due to (4.72) and due to the elementary inequality
$\sqrt{\cosh{\xi}} = \sqrt{2\cosh^2{(\xi/2)} - 1} \geq \sqrt{\cosh^2{(\xi/2)}} = \cosh{(\xi/2)}$,

$$\|\eta_{2,N}(\mathcal{F}, h)\| \leq \frac{C(\varphi, \alpha)h\|A^\alpha u_1\|}{2\pi} \sum_{k=N+1}^\infty e^{(x-1)\tilde{a}\sqrt{\cosh{(kh)}} - \alpha k h}$$

$$\leq \frac{C(\varphi, \alpha)h\|A^\alpha u_1\|}{2\pi} \sum_{k=N+1}^\infty e^{(x-1)\tilde{a}\cosh{(kh/2)} - \alpha k h} \tag{4.76}$$

$$\leq \frac{c\|A^\alpha u_1\|}{\alpha} \exp[(x-1)\tilde{a}\cosh{((N+1)h/2)} - \alpha(N+1)h],$$

where the constant c does not depend on h, N, x. Equating both exponentials in (4.75) and (4.76) for $x = 1$ by

$$\frac{\pi d}{h} = \alpha(N+1)h$$

we obtain for the step-size

$$h = \sqrt{\frac{\pi d}{\alpha(N+1)}}.$$

With this step-size, the error estimate

$$\|\eta_N(\mathcal{F}, h)\| \leq \frac{c}{\alpha}\exp\left(-\sqrt{\pi d\alpha(N+1)}\right)\|A^\alpha u_1\| \tag{4.77}$$

holds true with a constant c independent of x, N. In the case $x < 1$ the first summand in the exponent of $\exp[(x-1)\tilde{a}\cosh{((N+1)h/2)} - \alpha(N+1)h]$ in (4.76) contributes mainly to the error order. Setting $h = c_1 \ln N/N$ with a positive constant c_1, we remain asymptotically for a fixed x with an error

$$\|\eta_N(\mathcal{F}, h)\| \leq c\left[e^{-\pi dN/(c_1 \ln N)} + e^{-c_1(x-1)\tilde{a}N/2 - c_1\alpha \ln N}\right]\|A^\alpha u_1\|,$$

where c is a positive constant. Thus, we have proven the following result.

Theorem 4.8. *Let A be a densely defined strongly positive operator and $u_1 \in D(A^\alpha)$, $\alpha \in (0,1)$, then the Sinc-quadrature (4.73) represents an approximate solution of the homogeneous initial value problem (4.55) (i.e., $u(t) = u_{hl}(x) = E(x; A)u_1$) and possesses a uniform with respect to $x \in [0,1]$ exponential convergence rate with estimate (4.74) which is of the order $\mathcal{O}(e^{-c\sqrt{N}})$ uniformly in $x \in [0,1]$ provided that $h = \mathcal{O}(1/\sqrt{N})$ and of the order*

$$\mathcal{O}\left(\max\left\{e^{-\pi dN/(c_1 \ln N)},\ e^{-c_1(x-1)\tilde{a}N/2 - c_1\alpha \ln N}\right\}\right)$$

for each fixed $x \in [0,1]$ provided that $h = c_1 \ln N/N$.

Analogously we can construct the exponentially convergent approximation $u_{hr,N}(x)$ for $u_{hr}(x) = E(1 - x; \sqrt{A})u_0$ and the exponentially convergent approximation

$$u_{h,N}(x) = u_{hl,N}(x) + u_{hr,N}(x)$$

to the solution (4.58) of the homogeneous problem (4.56).

4.3.3 Inhomogeneous differential equations

In this section, we consider the inhomogeneous problem (4.56) whose solution is given by (4.57)- (4.60). We obtain an exponentially convergent approximation $u_{h,N}$ for the part $u_h(x)$ applying representation (4.58) and the discretization of the operator normalized hyperbolic sine family (4.73) described in the previous section.

To construct an exponentially convergent approximation $u_{p,N}$ to $u_p(x)$ we use the following Dunford-Cauchy representation through the Green function:

$$
\begin{aligned}
u_p(x) &= \int_0^1 G(x, s; A) f(s) ds \\
&= \int_0^1 \frac{1}{2\pi i} \int_{\Gamma_I} G(x, s; z) \left[(zI - A)^{-1} - \frac{1}{z} I \right] f(s) dz ds \qquad (4.78) \\
&= \frac{1}{2\pi i} \int_{-\infty}^{\infty} \mathcal{I}_p(x, \xi) d\xi
\end{aligned}
$$

with

$$\mathcal{I}_p(x, \xi) = \int_0^1 \mathcal{F}_p(x, s, \xi) f(s) ds,$$

$$\mathcal{F}_p(x, s, \xi) = G(x, s; \sqrt{z(\xi)})(a_I \sinh \xi - ib_I \cosh \xi) \left[(z(\xi) I - A)^{-1} - \frac{1}{z(\xi)} I \right],$$

$$G(x, s; z) \equiv G(x, s) = [\sqrt{z} \sinh \sqrt{z}]^{-1} \begin{cases} \sinh(x\sqrt{z}) \sinh((1 - s)\sqrt{z}) & x \le s, \\ \sinh(s\sqrt{z}) \sinh((1 - x)\sqrt{z}) & x \ge s. \end{cases}$$

Taking into account the last formula, we can represent (4.78) in the form

$$u_p(x) = \frac{1}{2\pi i} \int_{-\infty}^{\infty} [\mathcal{I}_{p,1}(x, \xi) + \mathcal{I}_{p,2}(x, \xi)] d\xi,$$

where

$$
\begin{aligned}
\mathcal{I}_{p,1}(x,\xi) &= \int_0^x \frac{\sinh\left(s\sqrt{z(\xi)}\right)\sinh\left((1-x)\sqrt{z(\xi)}\right)z'(\xi)}{\sqrt{z(\xi)}\sinh\left(\sqrt{z(\xi)}\right)} \\
&\quad \times \left[(z(\xi)I - A)^{-1} - \frac{1}{z(\xi)}I\right]f(s)ds, \\
\mathcal{I}_{p,2}(x,\xi) &= \int_x^1 \frac{\sinh\left(x\sqrt{z(\xi)}\right)\sinh\left((1-s)\sqrt{z(\xi)}\right)z'(\xi)}{\sqrt{z(\xi)}\sinh\left(\sqrt{z(\xi)}\right)} \\
&\quad \times \left[(z(\xi)I - A)^{-1} - \frac{1}{z(\xi)}I\right]f(s)ds.
\end{aligned}
\tag{4.79}
$$

Using the formula

$$
\sinh(\alpha)\sinh(\beta) = \frac{1}{2}\left[\cosh(\alpha+\beta) - \cosh(\alpha-\beta)\right],
$$

the fraction in the first formula in (4.79) can be represented as

$$
\begin{aligned}
&\frac{\sinh\left(s\sqrt{z(\xi)}\right)\sinh\left((1-x)\sqrt{z(\xi)}\right)}{\sinh\left(\sqrt{z(\xi)}\right)} \\
&= \frac{\cosh\left((1-x+s)\sqrt{z(\xi)}\right) - \cosh\left((1-x-s)\sqrt{z(\xi)}\right)}{2\sinh\left(\sqrt{z(\xi)}\right)}.
\end{aligned}
$$

Taking into account that $0 \le s \le x \le 1$ (it means that $x-s \ge 0$, $0 \le 1-x+s \le 1$) we deduce analogously to (4.67) that

$$
\begin{aligned}
\left|\frac{\cosh\left((1-x+s)\sqrt{z}\right)}{\sinh\left(\sqrt{z}\right)}\right| &= \left|\frac{e^{(1-x+s)\sqrt{z}} + e^{-(1-x+s)\sqrt{z}}}{e^{\sqrt{z}} - e^{-\sqrt{z}}}\right| \\
&= \left|\frac{e^{(1-x+s-1)\sqrt{z}} + e^{-(1-x+s+1)\sqrt{z}}}{1 - e^{-2\sqrt{z}}}\right| \\
&= \frac{e^{-(x-s)\sqrt{z}}\left|1 + e^{-2(1-x+s)\sqrt{z}}\right|}{\left|1 - e^{-2\sqrt{z}}\right|} \le \frac{2e^{-(x-s)\sqrt{z}}}{\left|1 - e^{-2\sqrt{z}}\right|} \\
&\le \frac{2e^{-(x-s)\sqrt{a_I\cosh(\xi)}}}{1 - e^{-2\sqrt{a_I}}},
\end{aligned}
\tag{4.80}
$$

$$
\begin{aligned}
\left|\frac{\cosh\left((1-x-s)\sqrt{z}\right)}{\sinh\left(\sqrt{z}\right)}\right| &= \frac{e^{-(x+s)\sqrt{z}} + e^{-(2-x-s)\sqrt{z}}}{\left|1 - e^{-2\sqrt{z}}\right|} \\
&\le \frac{e^{-(x-s)\sqrt{z}}\left(e^{-2s\sqrt{z}} + e^{-2(1-x)\sqrt{z}}\right)}{\left|1 - e^{(-2\sqrt{z})}\right|} \\
&\le \frac{2e^{-(x-s)\sqrt{a_I\cosh(\xi)}}}{1 - e^{-2\sqrt{a_I}}}.
\end{aligned}
\tag{4.81}
$$

In the case $0 \leq x \leq s \leq 1$ (it means that $s - x \geq 0$, $0 \leq 1 - s + x \leq 1$), we have

$$\frac{\sinh\left(x\sqrt{z(\xi)}\right)\sinh\left((1-s)\sqrt{z(\xi)}\right)}{\sinh\left(\sqrt{z(\xi)}\right)}$$

$$= \frac{\cosh\left((1-s+x)\sqrt{z(\xi)}\right) - \cosh\left((1-s-x)\sqrt{z(\xi)}\right)}{2\sinh\left(\sqrt{z(\xi)}\right)},$$

$$\left|\frac{\cosh\left((1-s+x)\sqrt{z}\right)}{\sinh\left(\sqrt{z}\right)}\right| \leq \frac{2e^{-(s-x)\sqrt{a_I \cosh(\xi)}}}{1 - e^{-2\sqrt{a_I}}}. \tag{4.82}$$

Estimates (2.32), (4.65), and (4.66) yield now the following inequalities for $\mathcal{I}_{p,1}(x,\xi)$, $\mathcal{I}_{p,2}(x,\xi)$:

$$\|\mathcal{I}_{p,1}(x,\xi)\| \leq \frac{2(1+M)Kb_I}{(1 - e^{-2\sqrt{a_I}})a_I^{3/2}} \int_0^x e^{-(x-s)\sqrt{a_I \cosh\xi} - 0.5|\xi|}\|f(s)\|ds,$$

$$\leq c\, e^{-0.5|\xi|} \int_0^1 \|f(s)\|ds,$$

$$\|\mathcal{I}_{p,2}(x,\xi)\| \leq \frac{2(1+M)Kb_I}{(1 - e^{-2\sqrt{a_I}})a_I^{3/2}} \int_x^1 e^{-(s-x)\sqrt{a_I \cosh\xi} - 0.5|\xi|}\|f(s)\|ds \tag{4.83}$$

$$\leq c\, e^{-0.5|\xi|} \int_0^1 \|f(s)\|ds,$$

where c is a constant independent of x,ξ. Analogously as above (see the proof of Theorem 4.8), one can also show for each x that $\mathcal{I}_{p,k}(x,w) \in \mathbf{H}^1(D_d)$, $0 < d < \varphi$, $k = 1, 2$ and

$$\|\mathcal{I}_{p,k}(x,\cdot)\|_{\mathbf{H}^1(D_d)} \leq c \int_0^1 \|f(s)\|ds \leq c \max_{s\in[0,1]}\|f(s)\|, \quad k = 1, 2$$

with a positive constant c depending on the spectral characteristics of A.

As the first step towards the full discretization we replace the integral in (4.78) by quadrature (4.73):

$$u_p(x) \approx u_{pa}(x) = \frac{h}{2\pi i} \sum_{k=-N}^{N} z'(kh)\left[(z(kh)I - A)^{-1} - \frac{1}{z(kh)}I\right]f_k(x) \tag{4.84}$$

where

$$f_k(x) = \int_0^1 G(x,s;z(kh))f(s)ds = f_{k,1}(x) + f_{k,2}(x), \quad k = -N, \ldots, N,$$

$$f_{k,1}(x) = \int_0^x \frac{\sinh((1-x)\sqrt{z(kh)})\sinh(s\sqrt{z(kh)})}{\sqrt{z(kh)}\sinh\sqrt{z(kh)}}f(s)ds, \tag{4.85}$$

and

$$f_{k,2}(x) = \int_x^1 \frac{\sinh(x\sqrt{z(kh)})\sinh((1-s)\sqrt{z(kh)})}{\sqrt{z(kh)}\sinh\sqrt{z(kh)}} f(s)ds. \qquad (4.86)$$

To construct an exponentially convergent quadrature for integral (4.85) we change the variables by

$$s = 0.5x(1 + \tanh\zeta) \qquad (4.87)$$

and obtain

$$f_{k,1}(x) = \int_{-\infty}^{\infty} \mathcal{F}_{k,1}(x,\zeta)d\zeta, \qquad (4.88)$$

instead of (4.85), where

$$\mathcal{F}_{k,1}(x,\zeta) = \frac{x\sinh((1-x)\sqrt{z(kh)})\sinh\left(x\sqrt{z(kh)}(1+\tanh\zeta)/2\right)}{2\sqrt{z(kh)}\sinh\sqrt{z(kh)}\cosh^2\zeta}$$
$$\times f\left(x(1+\tanh\zeta)/2\right).$$

We change the variables in integral (4.86) by

$$s = 0.5x(1 - \tanh(\zeta)) + 0.5(1 + \tanh(\zeta))$$

and obtain

$$f_{k,2}(x) = \int_{-\infty}^{\infty} \mathcal{F}_{k,2}(x,\zeta)d\zeta, \qquad (4.89)$$

instead of (4.86) with

$$\mathcal{F}_{k,2}(x,\zeta) = \frac{(1-x)\sinh(x\sqrt{z(kh)})\sinh\left(0.5(1-x)\sqrt{z(kh)}(1-\tanh(\zeta))\right)}{2\sqrt{z(kh)}\sinh\sqrt{z(kh)}\cosh^2\zeta}$$
$$\times f\left(0.5x(1-\tanh(\zeta)) + 0.5(1+\tanh(\zeta))\right).$$

Note that, with the complex variables $z = \zeta + i\nu$ and $w = u + iv$, equation (4.87) represents the conformal mapping $w = \psi(z) = x[1 + \tanh z]/2$, $z = \phi(w) = 0.5\ln\frac{w}{x-w}$ of the strip D_ν onto the domains $\mathcal{A}_\nu(x)$ (compare with the domain D_ν^2 in [62]) and $\mathcal{A}_\nu(x) \subseteq \mathcal{A}_\nu(1)$ $\forall x \in [0,1]$. It is easy to see that the images $\mathcal{A}_{1,\nu}(x)$ of the strip D_ν by the mapping $w = \psi_1(z) = 0.5x(1-\tanh(\zeta)) + 0.5(1+\tanh(\zeta))$, $z = \phi_1(w) = 0.5\ln\frac{x-w}{w-1}$ of the strip D_ν are all of the same form and are contained in $\mathcal{A}_\nu(1) = \mathcal{A}_{1,\nu}(0)$.

Due to (4.80)–(4.82) the integrands $\mathcal{F}_{k,1}(x,\zeta)$ and $\mathcal{F}_{k,2}(x,\zeta)$ satisfy on the real axis $\zeta \in \mathbb{R}$ (for each fixed $x \in [0,1]$) the estimates

$$\|\mathcal{F}_{k,1}(x,\zeta)\|$$
$$\leq \frac{x\exp\left[-(x - 0.5x(1+\tanh\zeta))\sqrt{a_I\cosh(kh)}\right]}{\left(1 - e^{-2\sqrt{a_I}}\right)|\sqrt{z(kh)}|\cosh^2\zeta}\|f(x(1+\tanh\zeta)/2)\|, \qquad (4.90)$$

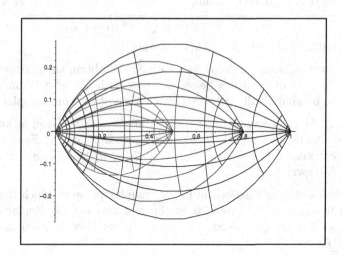

Figure 4.5: The images $\mathcal{A}_\nu(x)$ of the strip for $x = 0.5, 0.8, 1.0$, $\nu = 1$.

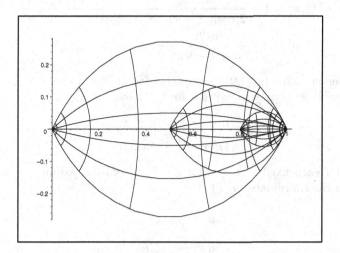

Figure 4.6: The images $\mathcal{A}_{1,\nu}(x)$ of the strip for $x = 0, 0.5, 0.8$, $\nu = 1$.

$\|\mathcal{F}_{k,2}(x,\zeta)\|$

$$\leq \frac{(1-x)\exp\left[-(0.5x(1-\tanh(\zeta))+0.5(1+\tanh(\zeta))-x)\sqrt{a_I\cosh(kh)}\right]}{\left(1-\mathrm{e}^{-2\sqrt{a_I}}\right)|\sqrt{z(kh)}|\cosh^2\zeta}$$

$$\times \|f(0.5x(1-\tanh(\zeta))+0.5(1+\tanh(\zeta)))\|,$$

which show their exponential decay as $\zeta \to \pm\infty$. To obtain the exponential convergence of the Sinc-quadratures below we show with the next lemma that the integrands can be analytically extended into a strip of the complex plane.

Lemma 4.9. *If the right-hand side $f(x)$ in (4.56) for $x \in [0,1]$ can be analytically extended into the domain $\mathcal{A}_\nu(1)$, then the integrands $\mathcal{F}_{k,1}(t,\zeta)$, $\mathcal{F}_{k,2}(t,\zeta)$ can be analytically extended into the strip D_{d_1}, $0 < d_1 < \pi/2$ and belong to the class $\mathbf{H}^1(D_{d_1})$ with respect to ζ.*

Proof. Let us investigate the domain in the complex plane in which the function $\mathcal{F}(x,\zeta)$ can be analytically extended to the real axis $\zeta \in \mathbb{R}$. Replacing in the integrand ζ to $\xi + i\nu$, $\xi \in (-\infty,\infty)$, $|\nu| < d_1$, we have in particular for the argument of f,

$$\tanh(\xi+i\nu) = \frac{\sinh\xi\cos\nu + i\cosh\xi\sin\nu}{\cosh\xi\cos\nu + i\sinh\xi\sin\nu} = \frac{\sinh(2\xi) + i\sin(2\nu)}{2(\cosh^2\xi - \sin^2\nu)}, \qquad (4.91)$$

$$1 \pm \tanh(\xi+i\nu) = q_r^\pm + iq_i^\pm$$

where

$$q_r^\pm(\xi,\nu) = 1 \pm \frac{\sinh 2\xi}{2(\cosh^2\xi - \sin^2\nu)} = \frac{\mathrm{e}^{\pm 2\xi} + \cos(2\nu)}{2(\cosh^2\xi - \sin^2\nu)},$$

$$q_i^\pm(\xi,\nu) = \pm \frac{\sin 2\nu}{2(\cosh^2\xi - \sin^2\nu)}.$$

The denominator in (4.91) is positive for all $\xi \in (-\infty,\infty)$ provided that $\nu \in (-\pi/2,\pi/2)$. It is easy to see that for $\xi \in (-\infty,\infty)$ we have

$$0 \leq q_r^\pm(\xi,\nu) \leq 2,$$
$$|q_i^\pm(\xi,\nu)| \leq |\tan\nu|.$$

Therefore, for each fixed x, ν and for $\xi \in (-\infty,\infty)$ the parametric curve $\Gamma_{\mathcal{A}}(t)$ given by (in the coordinates μ, η)

$$\mu = \frac{x}{2}q_r^-(\xi,\nu),$$

$$\eta = \frac{x}{2}q_i^-(\xi,\nu)$$

from (4.91) is closed and builds with the real axis at the origin the angle

$$\theta = \theta(\nu) = \arctan|\lim_{\xi\to\infty} q_i^-(\xi,\nu)/q_r^-(\xi,\nu)| = \arctan(\tan(2\nu)) = 2\nu.$$

For $\nu \in (-\pi/2, \pi/2)$ the domains $\mathcal{A}(x)$ for various $x \in [0,1]$ lie in the right half-plane ($q_r^{\pm}$ can not be negative) and fill the domain $\mathcal{A}_\nu(1)$ (see Fig. 4.5). Taking into account (4.70) and (4.90), we have

$$\|\mathcal{F}_{k,1}(x, \xi + i\nu)\|$$

$$\leq \left| \frac{x\, e^{-0.5x(1-\tanh(\xi+i\nu))\sqrt{a_I \cosh(kh)}}}{2(1 - e^{-2\sqrt{a_I}})|\sqrt{z(kh)}|(\cosh^2 \xi - \sin^2 \nu)} \right|$$

$$\times \|f(x(1 + \tanh(\xi + i\nu))/2)\|$$

$$\leq \left| \frac{x\, e^{-0.5x(q_r^- + iq_i^-)\sqrt{a_I \cosh(kh)}}}{2(1 - e^{-2\sqrt{a_I}})|\sqrt{z(kh)}|(\cosh^2 \xi - \sin^2 \nu)} \right| \|f(0.5x(q_r^+ + iq_i^+))\| \tag{4.92}$$

$$\leq \frac{x \exp\left(-0.5x\sqrt{a_I \cosh(kh)} \frac{e^{-2\xi} + \cos(2\nu)}{2(\cosh^2 \xi - \sin^2 \nu)}\right)}{2(1 - e^{-2\sqrt{a_I}})\sqrt{a_I}(\cosh^2 \xi - \sin^2 \nu)} \|f(0.5x(q_r^+ + iq_i^+))\|.$$

This inequality implies

$$\int_{-\infty}^{\infty} \|\mathcal{F}_{k,1}(x, \xi + i\nu)\| d\xi \leq c_1 \max_{w \in \mathcal{A}_\nu(1)} \|f(w)\| \int_{-\infty}^{\infty} \frac{x}{\cosh^2 \xi - \sin^2 \nu} d\xi$$

$$< c \max_{w \in \mathcal{A}_\nu(1)} \|f(w)\|.$$

This estimate yields $\mathcal{F}_{k,1}(x, w) \in \mathbf{H}^1(D_{d_1})$ with respect to w. The same conclusions are valid for $\mathcal{F}_{k,2}(x, w)$. $\qquad\square$

Under assumptions of Lemma 4.9, we can use the following Sinc quadrature rule to compute the integrals (4.88), (4.89) (see [62], p. 144):

$$f_k(x) = f_{k,1}(x) + f_{k,2}(x) \approx f_{k,N}(x) = f_{k,1,N}(x) + f_{k,2,N}(x)$$

$$= h \sum_{j=-N}^{N} \left[\mu_{k,1,j}(x) f(\omega_{1,j}(x)) + \mu_{k,2,j}(x) f(\omega_{2,j}(x)) \right], \tag{4.93}$$

where

$$\mu_{k,1,j}(x) = \frac{x \sinh((1-x)\sqrt{z(kh)}) \sinh\left(0.5x\sqrt{z(kh)}(1 + \tanh jh)\right)}{2\sqrt{z(kh)} \sinh \sqrt{z(kh)} \cosh^2 jh},$$

$$\omega_{1,j}(x) = 0.5x[1 + \tanh(jh)],$$

$$\mu_{k,2,j}(x) = \frac{(1-x)\sinh(x\sqrt{z(kh)}) \sinh\left(0.5(1-x)\sqrt{z(kh)}(1 - \tanh(jh))\right)}{2\sqrt{z(kh)} \sinh \sqrt{z(kh)} \cosh^2 jh},$$

$$\omega_{2,j}(x) = 0.5x[1 - \tanh(jh)] + 0.5[1 + \tanh(jh)],$$

$$h = \mathcal{O}(1/\sqrt{N}),$$

$$z(\xi) = a_I \cosh \xi - ib_I \sinh \xi.$$

Substituting (4.93) into (4.84) gives the following fully discrete algorithm to compute an approximation $u_{pa,N}(t)$ to $u_{pa}(t)$,

$$
u_{pa,N}(t) = \frac{h^2}{2\pi i} \sum_{k,j=-N}^{N} z'(kh)[(z(kh)I - A)^{-1} - \frac{1}{z(kh)}I]
$$
$$
\times [\mu_{k,1,j}(x)f(\omega_{1,j}(x)) + \mu_{k,2,j}(x)f(\omega_{2,j}(x))]. \tag{4.94}
$$

The next theorem characterizes the error of this algorithm.

Theorem 4.10. *Let A be a densely defined strongly positive operator with the spectral characterization a_0, φ and the right-hand side $f(x) \in D(A^\alpha)$, $\alpha > 0$ be analytically extended into the domain $\mathcal{A}_\nu(1)$, then algorithm (4.94) converges with the error estimate*

$$
\|\mathcal{E}_N(x)\| = \|u_p(x) - u_{ap,N}(x)\|
$$
$$
\leq \frac{c}{\alpha} e^{-c_1\sqrt{N}} \max_{w \in \mathcal{A}_\nu(1)} \|A^\alpha f(w)\| \tag{4.95}
$$

uniformly in $x \in [0,1]$ with positive constants c, c_1 depending on a_0, φ but independent of N.

Proof. We represent the error in the form

$$
\mathcal{E}_N(x) = u_p(x) - u_{pa,N}(x) = r_{1,N}(x) + r_{2,N}(x), \tag{4.96}
$$

where

$$
r_{1,N}(x) = u_p(x) - u_{pa}(x),
$$
$$
r_{2,N}(x) = u_{pa}(x) - u_{pa,N}(x).
$$

Taking into account estimate (4.83), we obtain for $h = \sqrt{2\pi d/(N+1)}$ similarly to as above (see estimate (4.77) in the proof of Theorem 4.8)

$$
\|r_{1,N}(x)\| = \|\frac{1}{2\pi i} \int_{-\infty}^{\infty} [\mathcal{I}_{p,1}(x,\xi) + \mathcal{I}_{p,2}(x,\xi)] \, d\xi
$$
$$
- \frac{h}{2\pi i} \sum_{k=-N}^{N} [\mathcal{I}_{p,1}(x,kh) + \mathcal{I}_{p,2}(x,kh)]\|
$$
$$
\leq c \exp\left(-\sqrt{0.5\pi d(N+1)}\right) \int_0^1 \|f(s)\| ds \tag{4.97}
$$
$$
\leq c \exp\left(-\sqrt{0.5\pi d(N+1)}\right) \max_{w \in \mathcal{A}_\nu(1)} \|A^\alpha f(w)\|.
$$

Due to (2.31), we have for the error $r_{2,N}(t)$,

$$
\begin{aligned}
\|r_{2,N}(t)\| &= \left\| \frac{h}{2\pi i} \sum_{k=-N}^{N} z'(kh)[(z(kh)I - A)^{-1} - \frac{1}{z(kh)}I]R_k(t) \right\| \\
&\leq \frac{h(1+M)K}{2\pi} \sum_{k=-N}^{N} \frac{|z'(kh)|}{|z(kh)|^{1+\alpha}} \|A^\alpha R_k(t)\|,
\end{aligned}
\tag{4.98}
$$

where $R_k(x) = f_k(x) - f_{k,N}(x)$,

$$
\begin{aligned}
\|A^\alpha R_k(t)\| &= \|A^\alpha(f_{k,1}(x) - f_{k,1,N}(x)) + A^\alpha(f_{k,2}(x) - f_{k,2,N}(x))\|, \\
&\leq \|A^\alpha(f_{k,1}(x) - f_{k,1,N}(x))\| + \|A^\alpha(f_{k,2}(x) - f_{k,2,N}(x))\|.
\end{aligned}
$$

The estimates (4.66), (4.80)-(4.82) analogously to (4.90) yield, for real ξ,

$$
\|A^\alpha \mathcal{F}_{k,1}(x,\xi)\| \leq c\, x e^{-2|\xi|} \|A^\alpha f((1+\tanh\xi)/2)\|.
$$

Due to Lemma 4.9, we have $A^\alpha \mathcal{F}_{k,1}(x,w) \in \mathbf{H}^1(D_{d_1})$, $0 < d_1 < \pi/2$ and we are in the situation analogous to that of Theorem 3.2.1, p. 144 from [62] with $A^\alpha f(w)$ instead of f which implies

$$
\begin{aligned}
&A^\alpha(f_{k,1}(x) - f_{k,1,N}(x))\| \\
&= \left\| \int_{-\infty}^{\infty} A^\alpha \mathcal{F}_{k,1}(x,\xi)d\xi - h \sum_{j=-\infty}^{\infty} A^\alpha \mathcal{F}_{k,1}(x,jh) + \|h \sum_{|j|>N} A^\alpha \mathcal{F}_{k,1}(x,jh) \right\| \\
&\leq \frac{e^{-\pi d_1/h}}{2\sinh(\pi d_1/h)} \|\mathcal{F}_{k,1}(x,w)\|_{\mathbf{H}^1(D_{d_1})} + h \sum_{|j|>N} c\, x e^{-2|jh|} \|A^\alpha f(x(1+\tanh jh)/2)\| \\
&\leq c e^{-2\pi d_1/h} \max_{w \in \mathcal{A}_\nu(1)} \|A^\alpha f(w)\| + hc\, x \max_{w \in \mathcal{A}_\nu(1)} \|A^\alpha f(w)\| \sum_{|j|>N} e^{-2|jh|},
\end{aligned}
$$

and, therefore, we have

$$
\|A^\alpha(f_{k,1}(x) - f_{k,1,N}(x))\| \leq c e^{-c_1\sqrt{N}} \max_{w \in \mathcal{A}_\nu(1)} \|A^\alpha f(w)\|,
\tag{4.99}
$$

where positive constants c, c_1 do not depend on x, N and k. The same is valid for $\|A^\alpha(f_{k,2}(x) - f_{k,2,N}(x))\|$, i.e., we have

$$
\|A^\alpha(f_{k,2}(x) - f_{k,2,N}(x))\| \leq c e^{-c_1\sqrt{N}} \max_{w \in \mathcal{A}_\nu(1)} \|A^\alpha f(w)\|.
$$

Now, estimate (4.98) can be continued as

$$
\begin{aligned}
&\|r_{2,N}(x)\| \\
&= \frac{h}{2\pi i} \sum_{k=-N}^{N} z'(kh)[(z(kh)I - A)^{-1} - \frac{1}{z(kh)}I]R_k(x) \leq c e^{-c_1\sqrt{N}} S_N
\end{aligned}
\tag{4.100}
$$

with

$$S_N = \sum_{k=-N}^{N} h \frac{|z'(kh)|}{|z(kh)|^{1+\alpha}}.$$

Using (4.65) and

$$|z(kh)| = \sqrt{a_I^2 \cosh^2(kh) + b_I^2 \sinh^2(kh)} \geq a_I \cosh(kh) \geq \frac{a_I e^{|kh|}}{2}$$

the last sum can be estimated by

$$|S_N| \leq \frac{c}{\sqrt{N}} \sum_{k=-N}^{N} e^{-\alpha|k/\sqrt{N}|} \leq c \int_{-\sqrt{N}}^{\sqrt{N}} e^{-\alpha t} dt \leq c/\alpha. \qquad (4.101)$$

Taking into account (4.99) and (4.101), we deduce, from (4.100),

$$\|r_{2,N}(x)\| \leq \frac{c}{\alpha} e^{-c_1 \sqrt{N}} \max_{w \in \mathcal{A}_\nu(1)} \|A^\alpha f(w)\|. \qquad (4.102)$$

The assertion of the theorem follows now from (4.96), (4.97), (4.102).   □

The exponentially convergent approximation to the solution of the inhomogeneous problem (4.56) is given by

$$u_N(x) = u_{h,N}(x) + u_{p,N}(x).$$

Example 4.11. We consider the inhomogeneous problem (4.56) with the operator A defined by

$$D(A) = \{u(x_1) \in H^2(0,1) : u(0) = u(1) = 0\},$$
$$Au = -u''(x_1) \, \forall u \in D(A).$$

The initial functions are $u_0 = u(0, x_1) = 0$, $u_1 = u(1, x_1) = 0$ and the right-hand side $f(x) = f(x; x_1)$ is given by

$$f(x; x_1) = 2\pi^2 \sin(\pi x) \sin(\pi x_1).$$

It is easy to see that the exact solution is $u(x) = u(x; x_1) = \sin(\pi x) \sin(\pi x_1)$. The algorithm (4.94) was implemented for $x = 1/2$, $x_1 = 1/2$ in Maple with Digits=16. Table 4.3 shows an exponential decay of the error $\varepsilon_N = |u(1/2, 1/2) - u_{pa,N}(1/2, 1/2)| = |u(1/2, 1/2) - u_N(1/2, 1/2)|$ with growing N.

N	ε_N
4	0.1872482412
8	0.829872855 e-1
16	0.115819650e-1
32	0.4730244e-3
64	0.46664e-5
128	0.63619e-9

Table 4.3: The error of algorithm (4.94) for $x = 1/2$, $x_1 = 1/2$.

Appendix: Tensor-product approximations of the operator exponential

We will show on the level of ideas that the operator exponential together with the tensor product of matrices allows us to make the computational costs of an approximation linear in d, at least, for some particular (but important) problems. As a simple example we consider the Dirichlet boundary value problem for the two-dimensional Laplacian in the unit square Ω with the boundary Γ:

$$-\Delta u \equiv -\frac{\partial^2 u}{\partial x_1^2} - \frac{\partial^2 u}{\partial x_2^2} = f(x_1, x_2), \ x = (x_1, x_2) \in \Omega,$$

$$u(x) = 0, \ x \in \Gamma.$$

Let $\omega_{j,h} = \{x_{1,i} = ih, \ i = 1, 2, \ldots, n\}, j = 1, 2$ be the equidistant grid with a step h in both directions and the operator of the second derivative in one direction be approximated by the well-known finite difference operator (in matrix form)

$$\left\{ \frac{d^2 u(x_i)}{dx^2} \right\}_{i=1,\ldots,n} \approx Du$$

with

$$u = \begin{pmatrix} u(x_1) \\ \cdot \\ \cdot \\ \cdot \\ u(x_n) \end{pmatrix}, \quad D = \frac{1}{h^2} \begin{pmatrix} -2 & 1 & 0 & 0 & \cdots & 0 \\ 1 & -2 & 1 & 0 & \cdots & 0 \\ 0 & 1 & -2 & 1 & \cdots & 0 \\ \cdot & \cdot & \cdot & \cdot & \cdot & \cdot \\ 0 & 0 & 0 & 0 & \cdots & 1 \end{pmatrix}.$$

If we provide the unknown grid function u_{ij} with two indexes and introduce the unknown matrix $U = \{u_{ij}\}_{i=1,\ldots,n;j=1,\ldots,n}$ (see Fig. A.1) and the given matrix $F = \{f(x_{1,i}, x_{2,j})\}_{i=1,\ldots,n;j=1,\ldots,n}$, then we obtain the approximate discrete problem

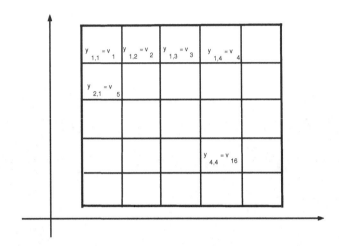

Figure A.1: To discretization of the Laplacian

$$DU + UD = -F. \tag{A.1}$$

This is a system of linear algebraic equations with $N = n^2$ unknowns which can be solved by the Gauss algorithm of the complexity $\mathcal{O}(n^6)$ (elementary multiplications). The use of the fast Fourier transform leads to an algorithm of complexity $\mathcal{O}(n^4 \ln n)$. Below we shall outline a scheme for obtaining an approximate algorithm of complexity $\mathcal{O}(n^2)$. The algorithm uses the exponential function.

Let us recall some facts from matrix theory [46]. Let $M_{m,n}$ be the space of matrices with m rows and n columns and with the usual operations, $M_{n,n} \equiv M_n$. If $A = \{a_{i,j}\} \in M_{m,n}$ and $B = \{b_{i,j}\} \in M_{p,q}$, then the tensor or Kronecker product $A \otimes B$ is defined as the block matrix from $M_{mp,nq}$,

$$A \otimes B = \begin{pmatrix} a_{11}B & a_{12}B & \cdots & a_{1n}B \\ a_{21}B & a_{22}B & \cdots & a_{2n}B \\ \cdot & \cdot & \cdot & \cdot \\ a_{m1}B & a_{m2}B & \cdots & a_{mn}B \end{pmatrix}.$$

It is easy to prove the following elementary properties of the tensor product:

$$\begin{aligned} (A \otimes B)(C \otimes D) &= AC \otimes BD, \\ (A \otimes B)^T &= A^T \otimes B^T, \\ (A \otimes B)^{-1} &= A^{-1} \otimes B^{-1}, \end{aligned} \tag{A.2}$$

provided that A and B are not singular in the last formula.

If A and B are matrices with complex elements, then $(\overline{A \otimes B}) = \overline{A} \otimes \overline{B}$ and $(A \otimes B)^* = A^* \otimes B^*$, where the operation $\overline{A}$ changes all elements of A to their complex conjugate elements and the operation A^* means the transposition and change to the complex conjugate elements, simultaneously.

If $A = \{a_{i,j}\} \in M_{p,q}$ and $B = \{b_{i,j}\} \in M_{s,t}$ then the direct sum $C = A \oplus B \in M_{p+s,q+t}$ is the block matrix

$$C = A \oplus B = \begin{pmatrix} A & 0_{p,t} \\ 0_{s,q} & B \end{pmatrix},$$

where $0_{k,l} \in M_{k,l}$ denotes the zero matrix, i.e., the matrix with all elements equal to zero. This operation is not commutative but associative. If $A \in M_{n,1}$ is an n-dimensional vector and $B \in M_{m,1}$ is an m-dimensional vector, then we define $C = A \oplus B \in M_{n+m,1}$ as an $n + m$-dimensional vector. If $A_i = \{a_i\}$ is a square (1×1)-matrix (i=1,...,m), then the direct sum $\bigoplus\limits_{i=1}^{m} A_i$ is the diagonal matrix $diag(a_1, \dots, a_m)$.

Let $A = \{A_{i,j}\}_{i=1,\dots,d}^{j=1,\dots,c}$ be a block matrix with elements A_{ij} being $(m_i \times n_j)$-matrices. Let $B = \{B_{i,j}\}_{i=1,\dots,c}^{j=1,\dots,f}$ be a block matrix with elements B_{ij} being $(n_i \times r_j)$-matrices. Then each product $A_{ij}B_{jl}$ is well defined and is an $m_i \times r_l$ -matrix. The product $C = AB \in M_{m,r}$ can be represented in block form $C = \{C_{il}\}_{i=1,\dots,d}^{l=1,\dots,f}$ with the elements $C_{il} = \sum\limits_{j=1}^{c} A_{ij}B_{jl}$, $i = 1, \dots, d$; $l = 1, \dots, f$.

Let $X = \{x_{i,j}\} \in M_{m,n}, A = \{a_{i,j}\} \in M_{p,m}, B = \{b_{i,j}\} \in M_{n,q}$, then the products $Y = AX, Z = XB, W = AXB$ are well defined. Let the mn-dimensional vector $x = \bigoplus\limits_{i=1}^{m} X^{(i)}$ be the direct sum of the matrix columns $X^{(i)}$. If we define the pn-dimensional vector $\bigoplus\limits_{t=1}^{n} Y^{(t)}$, then we have $y = \bigoplus\limits_{t=1}^{n} AX^{(t)} = (I_n \otimes A)x$ where I_n is the n-dimensional identity matrix.

Analogously, if $z = \bigoplus\limits_{t=1}^{q} Z^{(t)}$ then $Z^{(t)} = \sum\limits_{s=1}^{n} b_{st}X^{(s)} = \sum\limits_{s=1}^{n} (B^T)_{ts}I_m X^{(s)}$ where $(B^T)_{ts}$ is the (t, s)-element of the matrix B^T. Thus, we have

$$z = \bigoplus\limits_{t=1}^{q}\sum\limits_{s=1}^{n}(B^T)_{ts}I_m X^{(s)} = (B^T \otimes I_m)x \text{ and}$$

$$w = \bigoplus\limits_{t=1}^{q} W^{(t)} = (B^T \otimes I_m)(I_n \otimes A)x = (B^T I_n \otimes I_m A)x = (B^T \otimes A)x.$$

Using these formulas, the matrix equation

$$AX + XA^* = I_n$$

with square complex matrices A, X of the order n can be transformed to the form

$$(I_n \otimes A + \overline{A} \otimes I_n)x = e,$$

where $e = \bigoplus_{j=1}^{n} e_j$ and e_j is the j-th column of the matrix I_n. If $A \in M_n, B \in M_m$, then it follows from $[A \otimes I_m, I_n \otimes B] = (A \otimes I_m)(I_n \otimes B) - (I_n \otimes B)(A \otimes I_m) = A \otimes B - A \otimes B = 0$ that $A \otimes I_m$ and $I_n \otimes B$ are commutative.

The exponential function possesses some specific properties in connection with tensor operations. Actually, let $A \in M_n, B \in M_m$, then, due to the commutativity of the corresponding tensor products, we have

$$e^{A \otimes I_m + I_n \otimes B} = e^{A \otimes I_m} \cdot e^{I_n \otimes B}. \tag{A.3}$$

Furthermore, due to

$$
\begin{aligned}
e^{A \otimes I_m} &= \sum_{k=0}^{\infty} \frac{(A \otimes I_m)^k}{k!}, \\
e^{I_n \otimes B} &= \sum_{j=0}^{\infty} \frac{(I_n \otimes B)^j}{j!},
\end{aligned}
\tag{A.4}
$$

the arbitrary term in $e^{A \otimes I_m} e^{I_n \otimes B}$ is given by

$$\frac{(A \otimes I_m)^k (I_n \otimes B)^j}{k! j!}.$$

Imposing

$$
\begin{aligned}
(A \otimes I_m)^k (I_n \otimes B)^j &= (A^k \otimes I_m^k)(I_n^j \otimes B^j) = (A^k \otimes I_m)(I_n \otimes B^j) \\
&= (A^k I_n) \otimes (B^j I_m) = A^k \otimes B^j,
\end{aligned}
$$

we arrive at

$$\frac{(A \otimes I_m)^k (I_n \otimes B)^j}{k! j!} = \frac{A^k \otimes B^j}{k! j!},$$

and we have the arbitrary term of the tensor product

$$e^A \otimes e^B,$$

i.e., it holds that

$$e^{A \otimes I_m} e^{I_n \otimes B} = e^A \otimes e^B. \tag{A.5}$$

Returning to the example with the discretization of the Laplace equation, let us denote the i-th column of the unknown matrix U by u_i and of the given matrix F by f_i. Then in accordance with the rules of the tensor calculus, equation (A.1) can be written down as

$$\mathcal{L}_h u_h = f_h$$

with the block $(n^2 \times n^2)$-matrix $\mathcal{L}_h = \mathcal{L}_{1,h} + \mathcal{L}_{2,h}$,

$$\mathcal{L}_{1,h} = I_n \otimes D = \begin{pmatrix} D & 0 & 0 & 0 & \cdots & 0 \\ 0 & D & 0 & 0 & \cdots & 0 \\ 0 & 0 & D & 0 & \cdots & 0 \\ \cdot & \cdot & \cdot & \cdot & \cdots & \cdot \\ \cdot & \cdot & \cdot & \cdot & \cdots & D \end{pmatrix},$$

$$\mathcal{L}_{2,h} = D \otimes I_n = \frac{1}{h^2} \begin{pmatrix} -2I_n & I_n & 0 & 0 & \cdots & 0 \\ I_n & -2I_n & I_n & 0 & \cdots & 0 \\ 0 & I_n & -2I_n & I_n & \cdots & 0 \\ \cdot & \cdot & \cdot & \cdot & \cdots & \cdot \\ \cdot & \cdot & \cdot & \cdot & \cdots & I_n \end{pmatrix}$$

and the n^2-dimensional unknown vector u_h as well as the given vector f_h:

$$u_h = \begin{pmatrix} u_1 \\ u_2 \\ \cdot \\ \cdot \\ \cdot \\ u_n \end{pmatrix}, \quad f_h = \begin{pmatrix} f_1 \\ f_2 \\ \cdot \\ \cdot \\ \cdot \\ f_n \end{pmatrix}.$$

Note that the matrices $\mathcal{L}_{1,h}$ and $\mathcal{L}_{2,h}$ are commutative. Suppose we could approximate

$$\mathcal{L}_h^{-1} = \sum_{k=1}^{r} \alpha_k e^{\beta_k (I_n \otimes D + D \otimes I_n)}$$

with some α_k, β_k and a moderate constant r. Then using the properties (A.3) and (A.4) we would obtain an approximation

$$\mathcal{L}_h^{-1} = \sum_{k=1}^{r} \alpha_k e^{\beta_k D} \otimes e^{\beta_k D}.$$

Furthermore supposing that we could approximate $f_h = f_{1,h} \otimes f_{2,h}$, $f_{1,h} \in \mathbb{R}^n$, $f_{2,h} \in \mathbb{R}^n$ and using the property (A.2) give

$$u_h = \mathcal{L}_h^{-1} f_h \approx \sum_{k=1}^{r} \alpha_k (e^{\beta_k D} f_{1,h}) \otimes (e^{\beta_k D} f_{2,h}).$$

This is an algorithm of $\mathcal{O}(rn^2) = \mathcal{O}(n^2)$ complexity.

To solve a d-dimensional elliptic problem

$$\mathcal{L}u = f$$

in a d-dimensional rectangle one uses a discretization

$$\mathcal{L}_h u_h = (\mathcal{L}_{1,h} + \cdots + \mathcal{L}_{d,h}) u_h = f_h$$

with $N = n^d$ nodes (n nodes in each dimension, $h = 1/n$) and a matrix $\mathcal{L}_h \in \mathbb{R}^{n^d \times n^d}$. The complexity of the $u_h = \mathcal{L}_h^{-1} f_h$ computation is, in general, $\mathcal{O}(n^{2d})$ (again the curse of dimensional ity!). Supposing the tensor-product representations

$$\mathcal{L}_h^{-1} = \mathcal{L}_1 \otimes \mathcal{L}_2 \otimes \cdots \otimes \mathcal{L}_d, \quad f_h = f_1 \otimes f_2 \otimes \cdots \otimes f_d$$

we have, due to

$$\begin{aligned}
\mathcal{L}_h^{-1} f_h &= (\mathcal{L}_1 \otimes \mathcal{L}_2 \otimes \cdots \otimes \mathcal{L}_d) \cdot (f_1 \otimes f_2 \otimes \cdots \otimes f_d) \\
&= (\mathcal{L}_1 \cdot f_1) \otimes (\mathcal{L}_2 \cdot f_2) \otimes \cdots (\mathcal{L}_d \cdot f_d)
\end{aligned}$$

the polynomial in d of complexity $\mathcal{O}(dn^2)$.

The Kronecker product with the Kronecker rank r,

$$\mathcal{L}_h^{-1} = \sum_{k=1}^{r} c_k \mathcal{L}_1^{(k)} \otimes \mathcal{L}_2^{(k)} \otimes \cdots \otimes \mathcal{L}_d^{(k)},$$

does not influence this asymptotics.

To get a representation $\mathcal{L}_h^{-1} = \mathcal{L}_1 \otimes \mathcal{L}_2 \otimes \cdots \otimes \mathcal{L}_d$ the exponential function is helpful again, namely we can use the representations

$$\mathcal{L}_h^{-\sigma-1} = \frac{1}{\Gamma(\sigma+1)} \int_0^\infty t^\sigma e^{-t\mathcal{L}_h} dt, \sigma > -1,$$

or

$$\mathcal{L}_h^{-\sigma-1} = \int_{-\infty}^\infty e^{(\sigma-1)u - e^u \mathcal{L}_h} du, \quad \sigma > -1$$

with $\sigma = 0$.

These ideas for getting exponentially convergent algorithms of low complexity were implemented in [19] for various problems.

Bibliography

[1] W. Arendt, C.J.K. Batty, M. Hieber, and F. Neubrander, *Vector-valued laplace transforms and cauchy problems. monographs in mathematics, 96.*, Birkhäuser Verlag, Basel, 2001.

[2] D. C. Arov and I. P. Gavrilyuk, *A method for solving initial value problems for linear differential equations in hilbert space based on the cayley transform*, Numer. Funct. Anal. and Optimiz. **14** (1993), no. 5 & 6, 456 – 473.

[3] D.Z. Arov, I.P. Gavrilyuk, and V.L. Makarov, *Representation and approximation of the solution of an initial value problem for a first order differential equation with an unbounded operator coefficient in Hilbert space based on the Cayley transform*, Progress in partial differential equations (C. Bandle et al., ed.), vol. 1, Pitman Res. Notes Math. Sci., 1994, pp. 40–50.

[4] A. Ashyralyev and P. Sobolevskii, *Well-posedness of parabolic difference equations*, Birkhäeser Verlag, Basel, 1994.

[5] K.I. Babenko, *Foundations of the numerical analysis (in Russian)*, Nauka, Moscow, 1986.

[6] H. Bateman and A. Erdelyi, *Higher transcendental functions*, MC Graw-Hill Book Comp., Inc., New York, Toronto, London, 1988.

[7] C. Canuto, M.Y. Hussaini, A. Quarteroni, and T.A. Zang, *Spectral methods in fluid dynamics*, Springer-Verlag, Berlin, Heidelberg, New York et. all., 1988.

[8] T. Carleman, *Über die asymptotische Verteilung der Eigenwerte partieller Differentialgleichungen*, Ber. der Sächs. Akad Wiss. Leipzig, Math.-Nat.Kl. **88** (1936), 119–132.

[9] Ph. Clement, H.J.A.M. Heijmans, S. Angenent, C.J. van Duijn, and B. de Pagter, *One-parameter semigroups.*, CWI Monographs, 5. North-Holland Publishing Co., Amsterdam, 1987.

[10] Ju.L. Dalezkij and M.G. Krein, *Stability of solutions of differential equations in banach space (in Russian)*, Nauka, Moscow, 1970.

[11] Ronald A. DeVore and George G. Lorentz, *Constructive approximation*, Grundlehren der Mathematischen Wissenschaften [Fundamental Principles of Mathematical Sciences], vol. 303, Springer-Verlag, Berlin, 1993.

[12] M.L. Fernandez, Ch. Lubich, C. Palencia, and A. Schädle, *Fast Runge-Kutta approximation of inhomogeneous parabolic equations*, Numerische Mathematik (2005), 1–17.

[13] M. Filali and M. Moussi, *Non-autonomous inhomogeneous boundary cauchy problem on retarded equations*, Southwest Journal of pure and Applied Mathematics (2003), 26–37.

[14] H. Fujita, N. Saito, and T. Suzuki, *Operator theory and numerical methods*, Elsevier, Heidelberg, 2001.

[15] I.P. Gavrilyuk, *Strongly P-positive operators and explicit representation of the solutions of initial value problems for second order differential equations in Banach space*, Journ.of Math. Analysis and Appl. **236** (1999), 327–349.

[16] I.P. Gavrilyuk, W. Hackbusch, and B.N. Khoromskij, *H-matrix approximation for the operator exponential with applications*, Numer. Math. **92** (2002), 83–111.

[17] _____, *Data- sparse approximation to the operator-valued functions of elliptic operator*, Math. Comp. **73** (2004), 1297–1324.

[18] _____, *Data-sparse approximation of a class of operator-valued functions*, Math. Comp. **74** (2005), 681–708.

[19] _____, *Tensor-product approximation to elliptic and parabolic solution operators in higher dimensions*, Computing **74** (2005), 131–157.

[20] I.P. Gavrilyuk and V.L. Makarov, *Representation and approximation of the solution of an initial value problem for a first order differential equation in Banach space*, Z. Anal. Anwend.(ZAA) **15** (1996), 495–527.

[21] _____, *Explicit and approximate solutions of second order elliptic differential equations in Hilbert-. and Banach spaces*, Numer. Func. Anal. Optimiz. **20** (1999), no. 7&8, 695– 717.

[22] _____, *Explicit and approximate solutions of second-order evolution differential equations in hilbert space*, Numer. Methods Partial Differential Equations **15** (1999), no. 1, 111–131.

[23] _____, *Exponentially convergent parallel discretization methods for the first order evolution equations*, Computational Methods in Applied Mathematics (CMAM) **1** (2001), no. 4, 333– 355.

[24] _____, *Exponentially convergent parallel discretization methods for the first order differential equations*, Doklady of the Ukrainian Academy of Sciences (2002), no. 3, 1–6.

[25] _____, *Strongly positive operators and numerical algorithms without satura-tion of accuracy*, vol. 52, Proceedings of Institute of Mathematics of NAS of Ukraine. Mathematics and its Applications, Kyiv, 2004.

[26] _____, *Algorithms without accuracy saturation for evolution equations in Hilbert and Banach spaces*, Math. Comp. **74** (2005), 555–583.

[27] _____, *Exponentially convergent algorithms for the operator exponential with applications to inhomogeneous problems in Banach spaces*, SIAM Journal on Numerical Analysis **43** (2005), no. 5, 2144–2171.

[28] _____, *An explicit boundary integral representation of the solution of the two-dimensional heat equation and its discretization*, J. Integral Equations Appl. **12** (Spring 2000), no. 1, 63– 83.

[29] I.P. Gavrilyuk, V.L. Makarov, and V.L. Rjabichev, *A parallel method of high accuracy for the first order evolution equation in Hilbert and Banach space*, Computational Methods in Applied Mathematics (CMAM) **3** (2003), no. 1, 86– 115.

[30] I.P. Gavrilyuk, V.L. Makarov, and V. Vasylyk, *A new estimate of the Sinc method for linear parabolic problems including the initial point*, Computa-tional Methods in Applied Mathematics (CMAM) **4** (2004), 163– 179.

[31] J.A. Goldstein, *Semigroups of linear operators and applications*, Oxford Uni-versity Press and Clarendon Press, New York, Oxford, 1985.

[32] D. Henry, *Geometrical theory of semilinear parabolic equations*, Springer-Verlag, Berlin-Heidelberg-New York, 1981.

[33] T. Kato, *On linear differential equations in banach spaces*, Communications on pure and applied mathematics **IX** (1956), 479–486.

[34] M.A. Krasnosel'skii, G.M. Vainikko, P.P. Zabreiko, Ja.B. Rutitskii, and V.Ja. Stezenko, *Approximate solution of operator equations*, Wolters-Noordhoff, Groningen, 1972.

[35] M.A. Krasnosel'skij and P.E. Sobolevskij, *Fractional powers of operators act-ing in Banach spaces (in Russian)*, Doklady AN SSSR **129** (1959), no. 3, 499–502.

[36] M.A. Krasnosel'skij, G.M. Vainikko, P.P. Zabrejko, Ja.B. Rutizkij, and V.Ja. Stezenko, *Approximate solution of operator equations (in Russian)*, Nauka, Moscow, 1969.

[37] S.G. Krein, *Linear differential operators in banach spaces*, Nauka, Moscow, 1968 (in Russian).

[38] _____, *Linear differential operators in banach spaces*, Amer. Math. Soc., New York, 1971.

[39] Kwonk. and D. Sheen, *A parallel method for the numerical solution of integro-differential equation with positive memory*, Comput. Methods Appl. Mech. Engrg. **192** (2003), no. 41-42, 4641–4658.

[40] O.A. Ladyzhenskaja, V.A. Solonnikov, and N.N. Uralzeva, *Linear and quasilinear parabolic equations (in Russian)*, Nauka, Moscow, 1967.

[41] S. Larsson, V. Thomee, and L.B. Wahlbin, *Finite-element methods for a strongly damped wave equation*, IMA J. Numer. Anal. **11** (1991), 193–197.

[42] M. López-Fernández and C. Palencia, *On the numerical inversion of the laplace transform of certain holomorphic mappings*, Applied Numerical Mathematics **51** (2004), 289–303.

[43] M. López-Fernández, C. Palencia, and A. Schädle, *A spectral order method for inverting sectorial laplace transforms*, SIAM J. Numer. Anal. **44** (2006), 1332–1350.

[44] J. Lund and K.L. Bowers, *Sinc methods for quadrature and differential equations*, SIAM, Philadelphia, 1992.

[45] V. A. Marchenko, *Sturm-Liouville operators and applications. in: "operator theory: Advances and applications"*, vol. *22*, Birkhauser-Verlag, Basel, 1986.

[46] M. Marcus and H. Minc, *A survey of matrix theory and matrix inequalities*, Allyn and Bacon, Inc., Boston, 1964.

[47] G. Mastroianni and G.V. Milovanović, *Interpolation processes. basic theory and applications*, Springer, Berlin, Heidelberg, 2008.

[48] W. McLean and V. Thomee, *Time discretization of an evolution equation via laplace transforms*, IMA Journal of Numerical Analysis **24** (2004), 439–463.

[49] I.V. Melnikova and A.I. Filinkov, *An integrated family of m and n functions*, Dokl. Akad. Nauk **326** (1992), no. 1, 35–39 (in Russian).

[50] I.P. Natanson, *Constructive function theory*, Nauka, Moscow, 1949.

[51] ———, *Konstruktive Funktionentheorie*, Akademie-Verlag, Berlin, 1955.

[52] ———, *Constructive function theory. v. 1: Uniform approximation, translated from Russian N. Obolensky. v. 2: Approximation in mean, translated from Russian by John R. Schulenberger. v. 3: Interpolation and approximation quadratures, translated from Russian by John R. Schulenberger*, Ungar, New York, 1964, 1965, 1965.

[53] C. Palencia, *Backward Euler method for abstract time-dependent parabolic equations with variable domains*, Numer. Math. **82** (1999), 471– 490.

[54] A. Pazy, *Semigroups of linear operator and applications to partial differential equations*, Springer Verlag, New York, Berlin, Heidelberg, 1983.

[55] G.N. Polozhij, *Equations of mathematical physics (in Russian)*, Vysshaja shkola, Moscow, 1964.

[56] A.A. Samarskii, I.P. Gavrilyuk, and V.L. Makarov, *Stability and regularization of three-level difference schemes with unbounded operator coefficients in banach spaces*, SIAM Journal on Numerical Analysis **39** (2001), no. 2, 708–723.

[57] A.A. Samarskii, P.N. Vabischevich, and P.P. Matus, *Difference schemes with operator factors*, Kluwer Academic Publishers, Boston, Dordrecht, London, 2002.

[58] D. Sheen, I. H. Sloan, and V. Thomée, *A parallel method for time-discretization of parabolic problems based on contour integral representation and quadrature*, Math.Comp. **69** (2000), 177–195.

[59] _____, *A parallel method for time-discretization of parabolic equations based on laplace transformation and quadrature*, IMA Journal of Numerical Analysis **23** (2003), 269–299.

[60] M.Z. Solomjak, *Application of the semi-group theory to investigation of differential equations in Banach spaces (in Russian)*, Doklady AN SSSR **122** (1958), no. 5, 766–769.

[61] M Sova, *Problème de cauchy pour équations hyperboliques opérationnelles à coefficients constants non-bornés*, Ann. Scuola Norm. Sup. Pisa **3** (1968.), no. 22, 67–100.

[62] F. Stenger, *Numerical methods based on sinc and analytic functions*, Springer Verlag, New York, Berlin, Heidelberg, 1993.

[63] G. Szegö, *Orthogonal polynomials*, American Mathematical Society, New York, 1959.

[64] _____, *Orthogonal polynomials (with an introduction and a complement by J.L.Geronimus) (in Russian)*, State Publishing House of Physical and Mathematical Literature, Moscow, 1962.

[65] V. Thomée, *A high order parallel method for time discretization of parabolic type equations based on Laplace transformation and quadrature*, Int. J. Numer. Anal. Model. **2** (2005), 121–139.

[66] V. Thomee and L.B. Wahlbin, *Maximum-norm estimates for finite-element methods for a strongly damped wave equation*, BIT Numerical Mathematics **44** (2004), 165–179.

[67] H. Triebel, *Interpolation theory, function spaces, differential operators*, North-Holland, 1978.

[68] V. Vasylyk, *Uniform exponentially convergent method for the first order evolution equation with unbounded operator coefficient*, Journal of Numerical and Applied Mathematics (ISSN 0868-6912) **1** (2003), 99–104 (in Russian).

[69] ———, *Approximate solution of the cauchy problem for differential equation with sectorial operator in banach space (in ukrainian)*, Bulletin of Lviv University. Physical and Mathematical sciences **9** (2004), 34–46.

[70] J.A.C. Weideman and L.N. Trefethen, *Parabolic and hyperbolic contours for computing the bromwich integral*, Math. Comp. **76** (2007), no. 259, 1341–1356.

[71] Kōsaku Yosida, *Functional analysis*, Classics in Mathematics, Springer-Verlag, Berlin, 1995, Reprint of the sixth (1980) edition.

Index

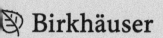

www.birkhauser-science.com

rontiers in Mathematics

his series is designed to be a repository for up-to-date research results which ave been prepared for a wider audience. Graduates and postgraduates as well as :ientists will benefit from the latest developments at the research frontiers in mathnatics and at the "frontiers" between mathematics and other fields like computer :ience, physics, biology, economics, finance, etc.

Dragović, V. / Radnović, M.,
ncelet Porisms and Beyond. Integrable .liards, Hyperelliptic Jacobians and Pencils of adrics (2011).
3N 978-3-0348-0014-3

e goal of the book is to present, in a com- ete and comprehensive way, areas of current search interlacing around the Poncelet porism: namics of integrable billiards, algebraic geo- etry of hyperelliptic Jacobians, and classical ojective geometry of pencils of quadrics. The ost important results and ideas, classical well as modern, connected to the Poncelet eorem are presented, together with a histori- l overview analyzing the classical ideas and eir natural generalizations. Special attention is id to the realization of the Griffiths and Harris ogramme about Poncelet-type problems and dition theorems. This programme, formulated ee decades ago, is aimed to understanding e higher-dimensional analogues of Poncelet oblems and the realization of the synthetic ap- oach of higher genus addition theorems.

Elworthy, K.D., Le Jan, Y., Li, X.-M.,
e Geometry of Filtering (2010).
3N 978-3-0346-0175-7

e geometry which is the topic of this book is t determined by a map of one space N onto other, M, mapping a diffusion process, or op- tor, on N to one on M.
tering theory is the science of obtaining or :imating information about a system from rtial and possibly flawed observations of it. The stem itself may be random, and the flaws in the servations can be caused by additional noise. :his volume the randomness and noises will be Gaussian white noise type so that the system

can be modelled by a diffusion process; that is it evolves continuously in time in a Markovian way, the future evolution depending only on the present situation.
We consider the geometry of this situation with special emphasis on situations of geometric, stochastic analytic, or filtering interest. The most well studied case is of one Brownian motion be- ing mapped to another with a consequent skew product decomposition (or equivalently the case of Riemannian submersions). This sort of decompo- sition is used to study in particular, classical filter- ing, (semi-)connections determined by stochastic flows, and generalised Weitzenbock formulae.

■ **Østvær, P.A.,** Homotopy Theory of C*-Algebras (2010). ISBN 978-3-0346-0564-9

Homotopy theory and C*-algebras are central top- ics in contemporary mathematics. This book intro- duces a modern homotopy theory for C*-algebras. One basic idea of the setup is to merge C*- algebras and spaces studied in algebraic topology into one category comprising C*-spaces. These objects are suitable fodder for standard homotopy theoretic moves, leading to unstable and stable model structures. With the foundations in place one is led to natural definitions of invariants for C*-spaces such as homology and cohomology theories, K-theory and zeta-functions.
The text is largely self-contained. It serves a wide audience of graduate students and researchers interested in C*-algebras, homotopy theory and applications.

■ **Borsuk, M.,** Transmission Problems for Elliptic Second-Order Equations in Non-Smooth Domains (2010).
ISBN 978-3-0346-0476-5